Trends and Challenges in Aerospace Offsets

Proceedings and Papers

CHARLES W. WESSNER, *Editor*

Board on Science, Technology, and Economic Policy

National Research Council

NATIONAL ACADEMY PRESS
Washington, D.C. 1999

NATIONAL ACADEMY PRESS • 2101 Constitution Ave., N.W. • Washington, DC 20418

Limited copies are available from:
Board on Science, Technology,
and Economic Policy
National Research Council
2101 Constitution Ave., N.W.
Washington, DC 20418
202-334-2200

Additional copies are available for sale from:
National Academy Press
Box 285
2101 Constitution Ave., N.W.
Washington, DC 20055
800-624-6242
202-334-3313 (in the Washington Metropolitan Area)

Library of Congress Catalog Card Number 98-89552
International Standard Book Number 0-309-06080-X

Printed in the United States of America

For the National Research Council, this project was overseen by the Board on Science, Technology, and Economic Policy (STEP), a new standing board of the NRC established by the National Academies of Sciences and Engineering and the Institute of Medicine in 1991. The mandate of the STEP Board is to integrate understanding of scientific, technological, and economic elements in the formulation of national policies to promote the economic well-being of the United States. A distinctive characteristic of STEP's approach is its frequent interactions with public and private sector decisionmakers. STEP bridges the disciplines of business management, engineering, economics, and the social sciences to bring diverse expertise to bear on pressing public policy questions. The members of the STEP Board and the NRC staff responsible for the project are listed below:

BOARD ON SCIENCE, TECHNOLOGY AND ECONOMIC POLICY

Steering Group on Policy Issues in Aerospace Offsets

Alan Wm. Wolff, *chair*
Managing Partner
Dewey Ballantine
Washington, D.C.

James F. Gibbons
Professor of Engineering
Stanford University
Stanford, Calif.

Dale Jorgenson
Frederic Eaton Abbe Professor of Economics
Harvard University
Cambridge, Mass.

Ruben Mettler
Chairman, CEO (ret.)
TRW, Inc.
Los Angeles

Mark B. Myers
Senior Vice President
Xerox Corporation
Stamford, Conn.

William J. Spencer
President and CEO
SEMATECH
Austin, Texas

Project Staff

Charles W. Wessner
Project Director

John C. Oldfield
Program Associate

Lena Lawrence
Administrative Associate

NATIONAL RESEARCH COUNCIL
BOARD ON SCIENCE, TECHNOLOGY, AND ECONOMIC POLICY

Sponsors

The National Research Council gratefully acknowledges the support of the following sponsors:

The White House National Economic Council

The Department of Commerce

Bureau of Export Administration
and
International Trade Administration

The Department of Defense

Office of the Deputy Under Secretary
for International and Commercial Programs

and

The Defense Advanced Research Projects Agency

Program Support for the Board on Science, Technology, and Economic Policy is provided by a grant from the Alfred P. Sloan Foundation

Contents

Preface

This volume is derived directly from the February 1997 request by the White House National Economic Council (NEC) to the National Research Council Board on Science, Technology and Economic Policy (STEP) to examine the impact of offsets on the U.S. aerospace industry. Specifically, the NEC asked that STEP organize a major workshop to examine the pressures facing U.S. companies to grant offsets in the increasingly competitive global aerospace markets for both defense and commercial aircraft and related products.[1] Although cognizant of the definitional and analytical challenges associated with this subject, the STEP Board accepted the NEC's request.

In accepting this task, the STEP Board's principal concern was that, in the time frame required for this initial effort, the Board would not be able to work through the analytical difficulties and overcome the data limitations associated with offsets. Consequently, the Board agreed to organize a comprehensive workshop and prepare a summary report of the workshop that would not, however, include recommendations or findings. The workshop was held at the National Academy of Sciences on June 9, 1997 and the

[1]The chapter entitled "Offsets" in: Trade Promotion Coordinating Committee, *National Export Strategy: Toward the Next American Century: A U.S. Strategic Response to Foreign Competitive Practices*. U.S. Government Printing Office, Washington, D.C., October 1996, describes offsets as compensation packages that are part of contract negotiations for large purchases such as aircraft. This description is elaborated in the issues paper in the appendix of the *National Export Strategy* report. However, it is important to keep in mind that different views on offsets sometimes lead to different definitions.

report, entitled *Policy Issues in Aerospace Offsets*, was published on June 30, 1997.[2]

The overarching objective of this first workshop was to provide a forum in which the various parties with an interest in aerospace offsets could come together to express their views on the practices, rationale, and current or future impact of offsets on U.S. national security, the competitiveness of the U.S. economy, especially the aerospace sector, and domestic employment in the aerospace industry. In his concluding remarks, Ambassador Alan Wm. Wolff, the workshop chair, observed that effective U.S. policy would require a broad consensus on the nature of the problems faced by the industry. This could lead to agreement on measures to maintain the domestic strength and international competitiveness of the U.S. aerospace sector. Continuing the dialogue would be a first step toward a consensus.

In an effort to maintain and deepen the dialogue opened at the first workshop, the responsible agencies agreed, in conjunction with representatives of aerospace labor and industry, to ask the STEP Board to convene a second meeting. In addition to continuing the dialog, this second meeting would also provide an opportunity for the STEP Board to have the interested parties review the papers it had commissioned to explore more rigorously the issues identified in the course of the first workshop.

Accordingly, a symposium was held at the National Academy of Sciences on January 14, 1998. Participants in the "Symposium on Trends and Challenges in Aerospace Offsets" considered the gathering a success, in part because the first workshop had encouraged the frank presentation of differing views and encouraged participants to take into account alternative perspectives, even if agreement was not reached on key points. At this symposium, the use of papers as a basis for discussion added nuance and texture to the presentations and helped identify points of broad agreement as well as issues that could benefit from further exploration.

Perhaps one of the most positive developments that emerged in the course of the 1998 symposium was the expansion of the terms of the debate beyond the scope, nature, and impact of offsets per se to broader issues of trade and investment and to policy issues such as streamlining adjustment assistance to displaced workers. Indeed, with regard to this latter point, many participants mentioned the need to simplify adjustment assistance programs for displaced workers. Other specific topics that emerged in the course of the symposium included calls for additional analysis of subsidies issues, for example, launch aids and export finance (including the adequacy of U.S. export finance), an assessment of the efficacy of the Multilateral Aircraft Agreement and the US-EU Bilateral Aircraft agreement and a range

[2]See Charles W. Wessner and Alan Wm. Wolff, eds., *Policy Issues in Aerospace Offsets* (National Academy Press, 1997).

of issues associated with market access. In addition there were several references to the need to review the adequacy of the U.S. investment in aerospace infrastructure, such as wind tunnels. Although these topics involve a broad range of policy questions, they do reflect recognition of the need to look at the U.S. aerospace industry, associated government policies, and the impact of foreign government policies as a whole.

As with the June 1997 workshop, the discussion recorded at the symposium did not address a cluster of issues sometimes associated with offsets. For example, domestic or foreign outsourcing decisions taken on the basis of commercial considerations remained outside the purview of the symposium, as were, at the other end of the spectrum, issues of arms control and proliferation. The focus of both meetings was commercial and military offsets, both direct and indirect, that companies are required to grant to complete sales of large systems. Several papers directed specific attention to the impact of offsets on key suppliers, sub-tier producers, and employment in this important part of the U.S. industrial base. This broad area of inquiry again proved to be more than adequate for the deliberations of a single meeting.

In sum, this symposium sought to achieve several interlocking objectives. Our first goal was to bring together a relatively small group of senior representatives from aerospace industry and labor in conjunction with academic experts to deepen our understanding of the offsets phenomenon and its consequences for the U.S. aerospace industry. Second, we sought to facilitate this discussion and ensure its rigor by presenting some of the best current analysis concerning the origins, causes, and current impact of offsets as well as future trends. Third, despite the different views expressed in the discussion, we sought to identify areas in which all parties agreed or partially agreed and to garner a wide range of potential policy recommendations to address the issues identified by participants. Last, in the course of the discussion every effort was made to place the offsets issue in the context of other developments and policy challenges having a major impact on the U.S. aerospace industry.

The presentations of speakers and participants were a challenge to summarize. Every effort has been made to capture the main points and supporting arguments of each speaker within the limitations of a summary report. We apologize in advance for any inadvertent errors or omissions in the summaries of the participants' presentations. However, the papers necessarily provide closer argumentation and include data and source material that has often been lacking in discussions of offsets. This and the willingness of the participants to expand the framework of the discussion, mark a significant advance in the national dialogue on aerospace offsets.

Charles W. Wessner

Acknowledgments

The Board on Science, Technology and Economic Policy would like to express its appreciation to the participants in this symposium for generously making available their time and expertise. The Board would particularly like to thank Dorothy Robyn, of the White House National Economic Council, for her leadership and support. The Board also wishes to express its appreciation to one of its members, Ambassador Alan Wm. Wolff for serving as chairman of this project. The Board would also like to recognize John Oldfield, a member of the STEP staff, for his contributions to the organization of the symposium and in the preparation of this volume. Lastly, the Board would like to express its appreciation to Dr. Charles Wessner, the project director, for bringing together informed representatives of labor, industry, government, and academia to discuss the impact of aerospace offsets and the challenges facing this important sector of the American economy.

A number of individuals deserve recognition for their willingness to review this report. These individuals were chosen for their diverse perspectives and technical expertise, in accordance with procedures approved by the NRC's Report Review Committee. The purpose of this independent review is to provide candid and critical comments that will assist the institution in making the published report as sound as possible and to ensure that the report meets institutional standards for objectivity, evidence, and responsiveness to the study charge. The review comments and draft manuscript remain confidential to protect the integrity of the deliberative process. We wish to thank the following individuals for their participation in the review process: Gerald Dinneen, the review coordinator; William J. Spencer, SEMATECH; Ruben Mettler, TRW, Inc.; Robert Scott, Economic

Policy Institute; Todd Watkins, Lehigh University; Joel Johnson, Aerospace Industries Association; Kenneth Flamm, the University of Texas at Austin and the Brookings Institution, and; John Mather, the National Academy of Sciences. Although these individuals have provided constructive comments and suggestions, it must be emphasized that responsibility for the final content of this report rests entirely with the STEP Board and the NRC.

I

INTRODUCTION

Introduction

The granting of offsets to support international exports of major aircraft systems has been a source of significant controversy. Critics of this practice believe that offsets undermine the U.S. manufacturing base, risk transferring commercial technology, some of which may also have national security implications, and result in the loss of high-value, high-wage jobs. Defenders of the practice argue that U.S. aerospace firms are driven by international competition to grant offsets. Suppliers must meet the demands of public or quasi-public purchasers for offsets which are normally a condition of increasingly important export sales. The practitioners also argue that they are exceedingly cautious in the types of commercially sensitive technology they transfer, not least because it is in their long-term self-interest to exercise such caution. They also suggest that the potential national security consequences are overstated and, in any event, subject to the rigorous U.S. export control system. With respect to the impact of offsets on employment, the firms responsible for implementing offsets agreements insist that U.S. domestic employment in the aerospace sector has been maintained—and indeed is growing again—precisely because they are able to strategically employ offsets to obtain sales in fiercely competitive global markets for major aerospace systems.

These markets are in fact large, and by general consensus, represent an essential source of sales for the U.S. aerospace industry. Aerospace itself has long been considered a strategic industry in terms of the sector's economic importance, military implications, and technological spillover. According to the Aerospace Industries Association (AIA), total U.S. aerospace sales in 1997 reached $130 billion, or about three percent of all U.S. industrial manufacturing activity. This figure includes some $50 billion in defense sales, of which $9.4 billion was exported. Overall, exports for the sector were quite significant. In 1997, the

AIA estimates exports at $50 billion, with imports of aerospace products reaching $16 billion. In 1997, both sales and employment were on the increase, with aerospace employment rising to approximately 870,000. The industry workforce has nonetheless declined substantially from its peak a decade ago when industry employment reached 1.35 million. Moreover, in late 1998, as this volume went to press, Boeing announced substantial cutbacks in its workforce.[1]

Information in support of the claims about the impact of offsets is difficult to obtain. Even though aerospace is considered a key U.S. industry, research on the impact of offsets is limited.[2] This symposium was designed to address the research gap while also ensuring that a wide range of viewpoints were considered. The summary of the symposium presentations and the vigorous discussion which accompanied them are supported by an exceptional series of commissioned papers. In addition to distinguished academic experts, representatives of labor and industry were asked to present their views for review and discussion. Together, the papers and proceedings offer a valuable opportunity to improve our understanding of current offset practices and associated policy issues.

The first challenge in any attempt to address the issues associated with offsets is definitional. As David Mowery notes in his introductory paper, "the definition of an offset is often very difficult . . ." not least because the "central defining characteristic of an offset is often in the eye of the beholder rather than in objective data or other indicators." Mowery adds that "another difficult issue in measurement of the magnitude and economic effects of offsets is the distinction between 'direct' and 'indirect' offsets." He defines direct offsets as transactions related to the specific product being exported, such as government to government coproduction agreements, licensed production offshore, or acquisition of components from suppliers in the purchasing country. Indirect offsets involve commitments by the exporter to either purchase products unrelated to the system being sold or to provide other forms of technical or commercial assistance to firms in the purchasing nation that are valued as some percentage of the export sale. Moreover, as Gordon Healey's presentation makes clear, the valuation of offsets

---

[1]The Boeing Company, which accounts for approximately 28 percent of the industry's employment, announced layoffs in December, 1998 on the order of 48,000, perhaps as high as 53,000, from a 1998 employment base of approximately 238,000. The cutbacks are attributed mainly to the effect of the Asian crisis on demand for Boeing aircraft. The cumulative impact of the Boeing cutbacks will be much larger, triggering substantial reductions in employment for the industry as a whole. Some analysts, such as EPI's Robert Scott, estimate the cumulative impact could be two additional aerospace job losses for each job eliminated at Boeing. Other participants in the symposium would point out, however, that the ability of U.S. firms to reduce employment in response to market forces is a competitive advantage for the U.S. economy. See "Boeing Raises Layoff Target to 53,000," by Elizabeth Douglass and Jeff Leeds, in *Los Angeles Times*, Wednesday, December 2, 1998.

[2]The bibliography in the Annex of this volume lists relevant work identified in the course of this project. The Bureau of Export Administration in the Department of Commerce maintains an active interest in the topic.

is itself an arbitrary process resulting from the perceived needs of the importing nation and the negotiating skill of the U.S. supplier. Consequently the measure and analysis of direct offsets is difficult. This difficulty is compounded in the case of indirect offsets.

Importantly, both representatives of aerospace labor and industry, as well as reputable analysts, recognize that offsets can result in loss of employment in some parts of the aerospace industry. For example, Mowery notes that anecdotal evidence suggests that there are negative consequences from offsets and similar transactions among the U.S. firms that supply the prime contractors. However, Mowery and other analysts emphasize that the impact of offsets is dwarfed by the much more significant consequences of the end of the Cold War and the concomitant restructuring of the industry. Moreover, representatives of industry would affirm that, from a national perspective, the sales made possible by offsets generate jobs or at least maintain employment throughout the U.S. aerospace supply chain. Representatives of labor and analysts such as the Economic Policy Institute's Robert Scott accept that offsets can contribute to maintaining employment among prime contractors but argue that offsets have also contributed to employment losses among the second- and third-tier of suppliers. Moreover, Scott is concerned that the long-term trends in U.S. aerospace employment are not positive. In his view, the combination of offsets, increased foreign competition, productivity growth, and other factors will all contribute to a decline in U.S. aerospace employment.

To meet this challenge and the challenge to the aerospace sector as a whole, Scott calls for a series of policy measures to restore and maintain the international competitiveness of the U.S. aerospace industry. Institutionally, this program would encompass an integrated, interagency task force lodged in the National Economic Council. Specific measures called for include regulatory reform, programs to stimulate aerospace research and products, and renewed efforts to negotiate improved international agreements with the European Union and to bring China into compliance with the GATT code. Scott emphasizes that the United States has in place a wide range of policies with direct impact on the aerospace industry but, unlike its competitors, the United States has yet to develop and implement "conscious and coherent" national policies for the aerospace industry.

Focusing on military offsets, Ken Flamm's analysis emphasizes the importance of understanding the economics of the aerospace sector, in particular the economies of scale required to absorb the large development costs characteristic of major weapons systems. With the post-Cold War decline in defense spending, exports have become critical, especially in Europe where national markets are smaller and defense procurement budgets have declined even further than in the United States. As a result, Flamm argues the European producers have a "driving need" to export, a need which they seek to meet by exporting systems with greater military capabilities. Ironically, these enhanced capabilities are themselves often acquired through U.S. cooperation with allies to develop capable, inter-operable weapons systems able to meet the requirements of coalition warfare. To meet this "capabilities competition," U.S. producers face strong incentives to provide similar capabilities, or see them acquired from a competing producer. The result Flamm suggests, is a situation where the United States cooperates technologically with its allies while competing with them for sales.

Flamm sees this dynamic leading either to a higher level of defense spending for the U.S., as it seeks to counter the enhanced foreign capabilities generated by export sales, or a more unsafe world for the United States. The preferable third option, he notes, would be to preserve cooperation on weapons development, while seeking international agreement to moderate the intense competition for defense sales.

From this perspective, the rationale for government interest in offset agreements turns on the asymmetry between the single purchaser (normally a foreign government) and the private companies supplying the aerospace products. Offsets are a means for the purchasing government to insert itself into private transactions for the benefit of national firms or to achieve other policy objectives. Flamm suggests there is a further complication in that foreign governments seek to encourage private U.S. firms to transfer technology often developed with U.S. taxpayer support. Flamm also points out that aerospace sales benefit from U.S. government support in other forms such as export licensing, government advocacy, and various financial subsidies, e.g., waivers of R&D recoupment. While these considerations are most evident with military sales, Flamm reminds us that these transactions represent a significant share of total aerospace sales. Moreover, even in the post-Cold War period, military R&D continues to play a "very important role" in, for example, the development of new jet engines. Flamm also differs with some analysts in finding that military cooperation has in fact had an impact in some countries, e.g. Japan, on the development of commercial aerospace technologies.

In his essay, Joel Johnson, the Vice President of the Aerospace Industries Association, agrees that many countries view aerospace both as a prestige industry and as a "technology driver" for the aerospace sector and for other leading high technology sectors such as electronics, advanced materials, and sensors. Moreover, because military aerospace acquisitions are made by governments with public funds, as are, de facto, many commercial acquisitions, there is a strong desire on the part of the purchasing government to link such acquisitions to domestic job creation and a strengthened national capability in aerospace and defense. Interestingly, the granting of offsets has the same rationale.

Despite these pressures, and the general trend toward internationalization, Johnson reminds the reader that U.S. exports of aerospace products remains strong. As noted in 1997, the U.S. exported $50 billion in aerospace products against $16 billion in imports, a performance which Johnson compares favorably with other U.S. high technology industries. Moreover, Johnson suggests that offsets must be seen in the context of market "internationalization," in which U.S. manufacturers seek to acquire more favorable financial participation, obtain new, improved technologies at lower risk and cost, and win better access to markets.

Notwithstanding the technological and export success of U.S. aerospace companies, Johnson joins several of the analysts here in recognizing the growing challenge posed to U.S. industry by Airbus and its would-be imitators on the commercial side and the increasingly stark competition for military markets. He sees the existing zero sum competition between U.S., European and Russian producers as being exacerbated by the desire of industrializing countries to capture a share of international military sales for their national economies.

Against this broader policy perspective, we add Todd Watkins' detailed review of the impact of strategic international sourcing in aircraft manufacturing on a U.S. aerospace supplier. Put in the context of the ongoing restructuring of the aircraft industry supply base, Watkins examines both the impact of demands for additional manufacturing responsibilities and risk sharing, and the encouragement of the internationalization of the supply base. The anonymous Generic Aircraft Manufacturing Company (GAMC) is a first-tier supplier to the major aircraft assemblers and a supplier of complex integrated assemblies. As such, GAMC must manage a network of more than 700 suppliers to which it increasingly delegates responsibility for a design, quality control, risk sharing and supplier management. Watkins suggests that the combination of increasing importance of foreign commercial markets and the implementation of collaborative supplier relationships poses significant dilemmas for major structural subcontractors such as GAMC and perhaps for all suppliers in the U.S. aircraft industry. In his view, the combination of market globalization and lean supplier development strategies encourages major subcontractors to outsource to the strongest foreign firms. Many of these firms have the strategy of acquiring technology abroad with the goal of independent production, a policy supported by the industrial policies of governments intent on developing their own aerospace industries.

While mindful of the limitations of a case study, Watkins concludes that the upper supplier tiers of the U.S. aircraft industry are being squeezed between the push towards lean practices and the pressures resulting from global sourcing on one hand and offset initiatives of top-tier companies on the other. He suggests that these pressures in combination with the developmental practices of a growing number of countries may pose serious challenges to the crucial middle-tier of the U.S. aircraft industry.

The essay by Owen Herrnstadt of the International Association of Machinists and Aerospace Workers, focuses on employment consequences of offsets. He argues that offset practices are increasing and that the government must play a central role in developing policies to address their negative impacts on U.S. aerospace workers. He joins Scott in observing that other aerospace nations have a coordinating body "charged with nurturing and advancing domestic aerospace manufacturing, technology acquisition and, of course, employment." To this end, he recommends the establishment of a formal commission to bring together representatives from industry, labor, government, and academia to facilitate the development of a national policy to foster the U.S. aerospace industry. Specifically, he recommends the commission should review the transfer of technology and employment, research and development, trade negotiations, export sales and financing, license production and co-production agreements, subcontractor production, countertrade, foreign investment, and labor adjustment programs. The commission could also serve as a means to better understand the competitive environment faced by U.S. companies and their workers, including offset requirements, and advise the government on appropriate policies.

The last paper in the volume by Gordon Healey, of the Defense Industry Offset Association, was made available after the meeting and therefore did not benefit from the same vigorous discussion as the other papers. It is, however, included in the annex because it provides an especially valuable perspective from the offset practitioner's point of view. Healey describes some of the definitional complexities surrounding aerospace offsets and adds support to Mowery's observation that the value of an offset is often in the eye of the beholder. He also emphasizes that U.S. aerospace companies do not voluntarily offer offsets to their customers. They see offsets as expensive, time consuming, difficult to manage, and politically unpalatable at home. Yet he sees offsets as a fact of life in the fierce competition for export sales, arguing that aerospace companies offer offsets with one goal, to win sales, and submits that without offsets, U.S. companies will lose sales.

From his practitioners' perspective, Healey suggests that solutions are likely to emerge only slowly as the industry evolves. In the meantime, the DIOA urges that no unilateral action be taken by the U.S. government, but does advocate improved data collection and multilateral discussions to limit offsets where possible. At the same time, he supports maintaining a domestic dialogue among concerned parties while urging that the impact of offsets be kept in perspective in light of the industry's performance.

Despite the complexity of the issues taken up by the symposium, it is our hope that the papers and discussions presented here will contribute to resolving questions concerning the definition, rationale, and consequences of offsets. At the same time, the analysis and discussion raise other questions concerning the future impact of offsets on the U.S. aerospace industry and its workers and, more broadly, the adequacy of current U.S. policy for the aerospace sector in an increasingly competitive global marketplace.

Charles W. Wessner

II

PROCEEDINGS

Welcome

Charles W. Wessner
National Research Council

Dr. Wessner opened the symposium by welcoming the participants and thanking them for their attendance. Noting that the meeting is intended to allow participants to explore important issues for the defense industrial base, long-term national security, and high-wage employment, he encouraged participants to ask questions at the end of the presentations. The symposium is designed to bring together a small, knowledgeable group to encourage a frank and open exchange. Although he emphasized the expectation that the courtesy that characterizes Academy deliberations would be respected, Dr. Wessner urged the participants to express their views freely on issues of importance to them. Garnering these views, having a frank and open discussion, and eliciting relevant questions are the goals of the symposium. Dr. Wessner then introduced Ambassador Alan Wm. Wolff, the symposium chair.

Symposium Introduction

Alan Wm. Wolff
Board on Science, Technology and Economic Policy

Ambassador Wolff welcomed the participants to the symposium on behalf of the Board on Science, Technology and Economic Policy (STEP). The STEP Board seeks to bring together experts in the disciplines of business management, engineering, and economics, with people with public policy experience, to address issues of national concern. In that regard, Ambassador Wolff welcomed the participation of senior public officials in today's symposium, including Dorothy Robyn of the White House National Economic Council, William Reinsch, Undersecretary of Commerce, and Page Hoeper, Deputy Undersecretary of Defense.

Turning to the topic of today's symposium, aerospace offsets, Ambassador Wolff observed that the aerospace sector of the U.S. economy remains a leading source of high-technology exports and exemplifies American technological preeminence. He also noted that today's symposium is designed as a sequel to the June 1997 workshop that, for the first time, brought together industry representatives directly involved in aerospace offsets to discuss the different types and rationales for offsets, as well as their consequences, with representatives of labor unions, expert academics, and government officials. The summary of that workshop, "*Policy Issues in Aerospace Offsets*," published shortly after the meeting, contains a rich discussion of the current policy issues relevant to this topic. This unique meeting also demonstrated the value of a balanced exchange of views to help policy makers understand the problems the industry faces in the fierce international competition for large contracts and follow-on work. The June meeting also led to a better understanding of some of the trends in demands for offsets. Questions were raised concerning the cumulative consequences of offsets, especially in conjunction with other industrial policy tools. In addition, the June discussion underscored the challenge the growth of offsets poses to the international trading system.

For the United States to meet these challenges, participants at the 1997 workshop suggested that a key first step would be to develop a better understanding of America's long-term economic interest. While the current financial crisis in Asia has led many U.S. journalists to believe that the Japanese model of economic development has been discredited, Ambassador Wolff doubted these views were shared by government agencies such as the Ministry for International Trade and Investment (MITI). Indeed the papers to be presented today suggest that the industrial policy plans of many countries are still an important factor in the aerospace industry.

To better understand these challenges and trends, there was a sense from the participants in the 1997 meeting that a more focused and detailed discussion would be of value. Today's symposium seeks to meet that need by attempting to further clarify the issues and long-term U.S. interests. To that end, the STEP Board commissioned a series of papers that are the basis of today's discussion. In reviewing the papers, Ambassador Wolff suggested that the participants might keep in mind the following questions:

- Do we have a clear idea of what is happening in aerospace offsets?
- How extensive are they, both in terms of dollar value and technological value?
- What is their impact, positive and negative, on both the industry and the country as a whole?
- Is there any part of the government that has a good understanding of the nature, content, and potential consequence of current offsets agreements?
- Does the government in fact have a need to know, or are offsets primarily a private matter even though public authorities are often involved as buyers?
- Do the demands for offsets that represent the objectives of a foreign government's industrial policy require a policy response by the United States? If so, what should be the main elements of that policy response?
- Are those foreign government industrial policies successful in their objective of transferring technology to better enable their companies to compete with U.S. firms?
- Are these policies succeeding cumulatively as the result of separate deals with competing vendors?
- Are these technologies being transferred crucial to long-term U.S. interests?

In summary, participants were asked to consider what is the environment in which U.S. companies are forced to compete, and what policy would be appropriate for the United States to pursue, both internationally and domestically.

Ambassador Wolff reminded participants that the goal of today's symposium is to be relevant to the current debate over U.S. policies. He encouraged a vigorous discussion, where it is recognized that all participants have a valid perspective. He asked participants to be specific but also to explore a broad range of policy options. These options could include, inter alia:

- multilateral or bilateral agreements to restrict or limit offsets;
- continued or enhanced government investment in research;
- the development of key infrastructure, such as wind tunnels;
- renewed attention to the national supply base, especially at the sub tier level—perhaps through a new manufacturing initiative directed at the needs of the aerospace industry;
- consideration of the impact of export control policy; and
- a review of the adequacy of export financing arrangements and international disciplines on export finance.

Ambassador Wolff closed by expressing the STEP Board's appreciation to the White House, to the Bureau of Export Administration and the International Trade Administration of the U.S. Department of Commerce, and to the Defense Advanced Research Projects Agency and the Office of the Deputy Undersecretary for International and Commercial Programs of the U.S. Department of Defense for their continued support of this program.

Opening Remarks

Dorothy Robyn
White House National Economic Council

Dr. Robyn began by thanking the National Academy of Sciences and the STEP Board for helping the administration think through this very complex and difficult issue. She then described two events that have influenced her thinking on the issue of aerospace offsets since the June conference.

The first event concerned the European review of the Boeing-McDonnell Douglas merger. This had been a difficult time in E.U.-U.S. bilateral relations. Indeed, although the subject was less threatening, the experience reminded her of the analysis of the Cuban missile crisis, where Graham Allison had used different conceptual models in his attempt to explain the Russians' behavior and the U.S.-Russian interaction. In this case, the U.S. decisionmakers were struggling to understand the Europeans' reaction to the merger.

The incident served as a strong reminder of the importance of the aerospace industry to both the United States and Europe. Based on recent experience, Dr. Robyn said that she finds it difficult to believe that the United States and Europe could come to an agreement to restrict the use of offsets. While it is an interesting intellectual suggestion that we are in a "prisoners' dilemma," with both sides increasing their use of offsets, it is hard to see how an agreement can be reached. Boeing and Airbus are now in direct competition, with Airbus taking market share directly away from Boeing rather than from McDonnell Douglas, as it has in the past. Nor is the playing field level. Airbus continues to develop new planes with government assistance, as allowed under the 1992 agreement. Given this high-stakes, zero-sum competition, it is hard to see how strategic outsourcing will not be seen as a critical competitive weapon by both parties.

The second event Dr. Robyn described was the failure of the Clinton administration to win fast-track trade authority. This loss was a stark reminder that the

United States may not be able to maintain the economic benefits of the international trading system, unless we preserve the underlying political consensus that supports that system. The discussion over offsets is one facet of that broader debate. Thus, it is important to have a rich and continuous discussion of offsets, and this is the important challenge facing this symposium.

Panel 1

Offsets in Commercial and Military Aerospace: An Overview

David C. Mowery
University of California, Berkeley

Dr. Mowery explained that his paper provides some background and an overview on aviation offsets. Defining offsets is a tricky issue, especially finding a definition that spans both military and commercial sales. On the military side, the classic definition of an offset is that it is a requirement for a transaction associated with a weapons system sale that goes beyond what is "normally" required to make the sale. Such a requirement may be a coproduction arrangement under a government-to-government agreement, licensed production offshore negotiated between the company and a foreign government, or purchase of components by the company for inclusion in the weapons system. Indirect offsets include the purchase of components, technology transfer, technical assistance, or other forms of assistance provided to the purchaser. That assistance could extend well beyond the aerospace industry. On the commercial side, the blurriness of the definition is even more pronounced, often involving some form of strategic outsourcing such as technology transfer, or purchasing of components that goes beyond what would normally be assumed to be commercially necessary. Given these definitional difficulties, it is subsequently hard to collect data on trends, magnitude, and content of offsets on the commercial side. It is also important to keep in mind the difference between direct and indirect offsets, especially because the impact of indirect offsets is likely to be more diffuse.

Military Offsets

The origins of military offsets go back to policies to support the rearmament and reconstruction of the defense industrial base and the development of interoperable weapons systems between the United States and the allies. Over time, a series of coproduction agreements evolved into initiatives by allies to become more involved in development, as well as licensed manufacturers. Data collected between 1980 and 1995 show no clear trend in the percentage of weapons systems sales accounted for by offsets, with a bump up to 80 percent in 1995 but long-term fluctuations around an average of 45–55 percent. There is a clearer trend, however, in the increase in the proportion of indirect to direct offsets. Only half of these indirect offsets involve aerospace; thus, an increasing number of military offsets affects companies outside the aerospace industry. This is especially true with respect to military exports to Pacific Rim countries.

There are complex interactions between commercial and military offsets. The many military coproduction and licensing agreements allowed the reconstruction of significant manufacturing capacity, employment bases, and aerospace infrastructures in countries allied with the United States. This in turn created a demand for offsets, much of which was sought in the form of civil or commercial business. Demand for commercial offsets began in a significant way in the 1960s and 1970s, a few decades after the beginning of military offsets.

Motives for Commercial Offsets

There are three motives driving commercial outsourcing:

- market access;
- a desire to spread risk and gain access to capital, including financing from governments interested in building their domestic aerospace industries; and
- access to product and process technological skills that are equal to or superior to that of U.S. domestic suppliers.

It is important to remember that these motives are reactions to both purely technical and commercial forces, and to trade-distorting government interventions. Separating these two factors, market forces and government intervention, is very difficult.

Dr. Mowery also noted that the U.S. firm that was the least successful in managing this international outsourcing and strategic alliance building was McDonnell Douglas. This inability to manage effectively these relationships, especially the inability to tap into funding sources to expand and maintain its product line, reduced McDonnell Douglas' share of business. Thus, there is a question as to whether aerospace companies can remain viable *without* maintaining these complex strategic alliance networks.

Impact of Offsets

Dr. Mowery observed that the effects of offsets are difficult to assess for three reasons:

- the lack of clear demarcations among the different categories;
- the long time required for effects to reveal themselves; and
- the difficulty in specifying the counterfactual (i.e., what would have happened in the absence of the offset).

Thus, a key analytical issue is whether U.S. firms could succeed in making substantial international sales without employing offsets. Answering this counterfactual is very difficult. However, it is clear that, absent some level of offsets, foreign sales are not likely to be made.

A second issue is the impact of indirect versus direct offsets on technology transfer and employment in the aerospace industry. The impact of indirect offsets, which are the more rapidly growing portion of offsets, is much more difficult to trace.

A third question on the commercial side concerns our ability to tease out what portion of the offsets are in response to government pressure.

Another issue concerns the outsourcing activities of non-U.S. companies. Dr. Mowery suggested that there may be a broader pattern of increased intra-industry trade in components and subassemblies, which may result in increased exports by U.S. suppliers. He noted that his paper shows that the trade surplus in U.S. aircraft parts has, in fact, increased in the past eight years.

These questions are especially relevant when looking at the employment impact, particularly the comparison of the impact from indirect and direct offsets. This type of analysis requires a complex set of distinctions between the wages and the relative labor content of the nonaerospace industries affected by indirect offsets versus the aerospace industry affected by direct offsets and the export sale.

Impact of Technology Transfer

Concerning the impact of technology transfer, Dr. Mowery noted that there is little evidence that technology transfer through military offsets has had a major impact on building up the civilian aerospace production and design capabilities of non-U.S. firms. Technology transfer *has* expanded the capacity and manufacturing capabilities of non-U.S. aerospace industries. However, if we take into account that, particularly in Western Europe, there was an aerospace industry that began the post-World War II period at a high-level of technological capability, it is not clear that the offset transactions, per se, have had a major effect on building up the prime contractor tier of non-U.S. aerospace industries. Still, it remains difficult to trace the impact of these technology transfer agreements for long periods of time. Certainly the generic capabilities, such as production capacity, tooling, and the maintenance of a skilled work force, have been strengthened by some of these agreements. Nevertheless, entry to the prime contractor tier, especially in the commercial sector, remains difficult. The trend in the industry has been the exit of prime contractors, not entry.

Consequently, Dr. Mowery suggested that the most likely impacts of offsets will be on the supplier tier. Because of the many influences affecting this sector of the industry, it is difficult to analyze the impacts of offsets. It is a large segment of the industry with many small firms and has not been very well tracked by public data.

The Industry Outlook

Concerning the trends and outlook for the industry, the most rapidly growing markets for commercial aircraft are outside the United States. However, the largest single market will continue to be North America, where demands for offsets are more manageable. Over a longer period, continued bilateral competition between Boeing and Airbus, and trilateral competition among Rolls-Royce, Pratt & Whitney, and General Electric, will continue the pressure for offset agreements. On the military side, substantial excess capacity means that competition in weapons systems is likely continue to intensify.

A Limited Role for Offsets

Dr. Mowery stressed that the contribution of offsets to employment shifts, compared with the broader trends of internationalization in the aerospace industry, is quite modest. The overall technological and employment impact of offsets on the aerospace industry is small when compared with the impacts of rationalization and defense downsizing. Thus, although not a trivial issue, the employment impacts of offsets per se are not the most important problem.

Nonetheless, there remains a legitimate basis for developing a policy response to displacement and employment loss in this industry because of increased internationalization—as there is to respond to trade-related displacement in any industry. Any response to offsets and broader trends of aerospace restructuring needs to deal with the trade-distorting policies of other governments, combined with ways of dealing with the adjustment issue domestically.

Policy Tools

The United States has a number of international tools to deal with trade-distorting behavior of other governments in both the commercial and the military sectors. These include

- international codes of conduct for military sales;
- various trade agreements under the General Agreement on Tariffs and Trade (GATT) such as government procurement and trade-related investment measures, although national security exemptions generally preclude their applicability to military offsets;

- various agreements on trade on large civil aircraft, such as the United States–European Union agreement;
- prohibitions on offsets in U.S. military sales financed under the foreign military sales funds program; and
- bringing Pacific Rim nations into negotiations over codes of conduct for military sales and elevating the offsets issue in the negotiations with the People's Republic of China on GATT accession.

However, it is important not to let offsets dominate discussion over the very real costs and adjustment issues associated with increased internationalization in the aerospace industry. The costs of offsets, both economically and politically, are very real. But analysts and policy makers must keep a sense of proportion in addressing the different causes and fashion the tools to deal with what are different sets of problems.

DISCUSSANTS

Sally Bath
U.S. Department of Commerce

Although agreeing with Dr. Mowery's description of the growing internationalization of the commercial aerospace industry, Ms. Bath observed that, at the same time, there are continuing efforts toward greater internalization now under way in Europe. The appearance of new European government subsidies, specifically designed to displace U.S. supplier firms from the European market, is particularly disconcerting. These targeted subsidies may have a larger impact on U.S. employment than offset agreements and may represent a first salvo against the U.S. subcontractor industry. She suggested that this is especially worrisome coming at a time, as Dr. Mowery's paper points out, when U.S. parts exports are increasing. Dealing with this emerging threat will require concentrated negotiations and discussions with our European aerospace colleagues.

Consequently, in Ms. Bath's view, the two trends of internationalization and an increased threat from European subsidies must be carefully watched. Although internationalization presumably cannot be stopped, it has not had that much of an impact on employment. The worldwide defense decline has been a greater factor. Changing production technology, which requires new skills and, in some cases, fewer people in the production process, is also a major factor. These changes in technology and production processes have not been looked at adequately. If we cannot maintain the flexibility to address quickly and adequately the requirements of technology change, we are damaging our long-term competitive position. The industry must retain the flexibility to adopt new processes, regardless of the impact on employment, and have the ability to quickly apply new technologies, which will often require new skills.

Steve Beckman
United Auto Workers

Mr. Beckman began by pointing out that the United Auto Workers (UAW) interest in this issue is both from the standpoint of a union that directly represents workers in the industry and from a concern over broader trade policy, especially given the previous internationalization of the auto industry. Although the experience of internationalization in the auto industry is no way identical to that of the aerospace industry, there are some common themes.

With respect to the policy portion of Dr. Mowery's paper, there appear to be three main sections: corporate policies, trade and international policies, and adjustment policies. In the sections on corporate policies and trade policies, the paper seems to say that the status quo is fine. To Mr. Beckman, the status quo is a serious problem for workers in the industry, and for the national interest. He disagrees with the proposition that there is no need for a business–labor–government debate or discussion on internationalization because the government has no role to play. On the contrary, in light of the industry's military connections and the fact that many countries have targeted aerospace as a lead economic development priority, there is an appropriate government role. National security and economic development concerns, including technology, jobs, and incomes, are huge issues of national interest.

National Decision Making at the Firm Level

Mr. Beckman remarked that the status quo is to allow companies to decide what form of internationalization will take place, what the response will be to foreign governments' policies, and how the development of the U.S. industry will proceed. However, there is necessarily a national interest in these questions, not just a corporate interest. This raises the issue of the identity between what is in a company's best interest and what is in a nation's best interest.

It is both necessary and useful to have consultation and information from the involved companies, to provide advice, counsel, and guidance on how the government should consider the national interest. As has been mentioned several times, the information needed to encourage rigorous analysis and thoughtful decision making is absent. There needs to be a mechanism for the exchange of information. Without such, we will continue with the status quo, with companies continuing to make decisions on behalf of the national interest and the government unable to assess adequately the consequences.

International Negotiations: Modest Accomplishments

Concerning trade and international policies, Mr. Beckman noted that there are numerous ongoing or potential negotiations. However, the accomplishments to date are incredibly modest. The U.S. government is able to negotiate agreements reducing trade barriers; yet the ability of those agreements to actually accomplish anything in the marketplace is not very well demonstrated. Our history of negotiations with Japan underscores this problem, as it appears will be the case with our negotiations with China. As mentioned in Dr. Mowery's paper, China's entry into the World Trade Organization will likely allow a 10–15-year period of transition. Much can be accomplished by China during this time, creating a major enforcement problem later on.

Adjustment Policy

This leaves adjustment policy. Mr. Beckman said that, although most people agree in principle concerning the importance of adjustment assistance, resources for such programs remain scarce. Companies are unwilling to fund adjustment assistance themselves; nor are they willing to pay higher taxes to fund government programs. In addition to a lack of funds, procedures for workers to qualify have always been onerous. Taking issue with the central premise of adjustment, that overall the economy will benefit from the displacement, Mr. Beckman argued that, in the case of aerospace, this is an industry that generates high value-added. Moving employment and production out of such a high value-added industry may not benefit the economy, especially if the movement is to lower value-added industries. He points out that Dr. Mowery's paper discusses imports of avionics from Canada displacing imports of potash. In his view, moving workers from the aerospace sector to potash does not appear to be a net benefit to the U.S. economy.

The key question is why adjustment is always left up to the individual worker. Beckman asked whether a European Commission employee would ever say that societies and workers must adjust to the technology, regardless of the impact on employment. The European attitude, and the attitude of much of the rest of the world, is very different than that of the United States regarding trade and employment. This is something that neither our trade negotiators nor the public appreciates.

The potash example brings to line the famous comment that there is no difference for the U.S. economy between potato chips and computer chips. That comment implies that it does not matter what the United States produces. Mr. Beckman stated that the view of the UAW is that, what is produced in United States matters very much—both for current employment and future economic development. This viewpoint is shared by policy makers around the world.[1]

[1]For a discussion of these perspectives, and the programs that result, see *Conflict and Cooperation in National Competition for High-Technology Industry* (National Academy Press, Washington, D.C., 1998, pp. 12–40). See especially Box A, "Why Are Countries Concerned about Their High-Technology Industries," pp. 33–35 and Box H, "The Stakes in Aerospace Competition," pp. 88–90.

Assessing both Exports and Imports

Mr. Beckman closed by remarking that both Dr. Mowery and Ms. Bath had commented on the exports of aerospace parts to Europe as an indicator of how well the industry was doing. Mr. Beckman argued that only looking at exports does not give a clear picture as to the employment impacts of trade. The effect of imports must also be taken into consideration. The overall balance of trade must be looked at when assessing employment effects. Ms. Bath objected, arguing that she is one of the few analysts in the U.S. government who looks at both imports and exports. Dr. Mowery also pointed to the use of both export and import data in the paper.

GENERAL DISCUSSION

Alan Wm. Wolff, Dewey Ballantine and STEP Board: Ambassador Wolff opened the general discussion by commenting on the lack of information with respect to foreign governments' industrial policy plans. He cited his recent experience trying to gain information on China's five-year plan for industry. He agreed that other forces may swamp the impact of foreign government plans on the U.S. aerospace industry. But without access to the information, it is very difficult to make that determination.

Steve Sleigh, International Association of Machinists and Aerospace Workers: Mr. Sleigh raised the question as to Dr. Robyn's comment in the opening remarks that there is no alternative to strategic outsourcing, and to Dr. Mowery's assertion that McDonnell Douglas failed because of its lack of strategic outsourcing. He pointed out that McDonnell Douglas planes contained more outsourced parts than either Boeing or Airbus.

Kirk Bozdogan, Massachusetts Institute of Technology: Dr. Bozdogan also questioned the comment that McDonnell Douglas' demise was due to the failure to establish strategic alliances. On the commercial side, McDonnell Douglas did a good job in integrating its suppliers into its production, design, and development process. He then raised the question about exports of aerospace parts and components. Because much of the worldwide installed base of aircraft is U.S. made—either Boeing or McDonnell Douglas—most of the spare parts to sustain these aircraft must naturally come from the United States. However, the U.S. aerospace industry cannot survive by simply selling spare parts.

John Sandford, Rolls-Royce, N.A.: Noting that his perspective is that of the CEO of an aerospace company, Mr. Sandford pointed out that in the commercial sector there has been a massive shift in the reason for outsourcing. It is no longer simply for market access. Access to capital is now the major reason behind strategic outsourcing. Cash flow considerations are very important, and margins within the industry are very poor. Risk sharing and a steady stream of payments made for work done in-country are important considerations in placing production. The investment made by foreign companies, through their governments, in the development of U.S. products is very large. That trend will continue in the future.

Corporate versus Government Balance Sheets

Mr. Sandford emphasized that one of the challenges we face is that decision makers in this process are using the balance sheet of corporations rather than the balance sheet of governments. Thus, European governments can afford to invest in their own products and in their industry to gain a share of the U.S. defense business. He urged analysts to "follow the money."

Randy Barber, Center for Economic Organizing: Mr. Barber had a question concerning Dr. Robyn's comments. He asked if she was describing a return to the competitive situation in commercial aircraft similar to that in the 1970s. At that time, there was fierce competition with high-level government intervention on the European side.

In addition, he thought that Dr. Robyn stated that it was inconceivable that strategic outsourcing would not be part of Boeing's response. This assumes that direct government assistance, similar to that of the Europeans, is not an option in the United States. This situation would suggest that Boeing has become the U.S. national champion, but without a corresponding national responsibility.

Dorothy Robyn, White House National Economic Council: Dr. Robyn clarified that she was simply referring to the fact that Boeing and Airbus are now competing directly for market share. Whereas in the past Airbus took market share away from McDonnell Douglas, any increase in Airbus's market share will now come directly at the expense of Boeing.

She added that there is a debate as to whether and to what extent government actions such as procurement and research and development (R&D) support are actually helpful to the industry. The United States has never provided direct aid to Boeing. The Europeans argue that U.S. defense procurement and R&D provide much more benefit than the United States believes. In fact, the U.S. Defense Department (DoD) now believes the reverse, that defense benefits more from expertise gained in the commercial sector. Dr. Robyn noted that her point is simply that Boeing must compete with a company that is still in the position to receive large direct subsidies.

She also pointed out, as is discussed in Dr. Scott's paper, that Airbus also does a great deal of strategic outsourcing. This has made them more competitive. In this environment of heightened competition, it seems naive to think that Boeing will do less strategic outsourcing.

David Welch, Defense Industry Offset Association and Lockheed Martin: Mr. Welch clarified some details regarding one of the examples of an offset deal in

Dr. Mowery's paper and pointed out the positive employment benefits of that deal. He urged participants to be very careful about advocating any legislation that could end up hurting export sales.

Adjustment Assistance

Owen Herrnstadt, International Association of Machinists and Aerospace Workers: Mr. Herrnstadt agreed with Dr. Mowery about the confusion over direct and indirect offsets and about the lack of data. Given these difficulties, he questioned the conclusion in Dr. Mowery's paper that the employment impact of offsets is negligible. Without adequate data, such a conclusion is both hard to reach and hard to accept. Mr. Herrnstadt also pointed out that, although the paper called for improved adjustment assistance, one of the difficulties in receiving assistance is proving the cause of the displacement. This also points out the need for better information and data to inform the policy debate.

David Mowery, University of California, Berkeley: Responding to a number of comments made during the discussion, Dr. Mowery emphasized that the employment effects of offsets are only negligible with respect to other forces affecting the industry. This finding is broadly consistent with findings of others, including those in Dr. Scott's paper that will be discussed later in the day. With respect to adjustment assistance, Dr. Mowery offered that he has always believed that adjustment assistance should be made available, *regardless* of the cause of the displacement. Creation of separate programs based on the cause of displacement is poor policy design and needlessly adds administrative costs.

He reiterated his belief that the decline and fall of McDonnell Douglas was due to its inability to develop new products, which was due, in part, to its inability to get risk-sharing partners on the commercial side of the business. Although there may not be a direct causal effect, there is a connection.

He also pointed out that any discussion on offsets should not be dominated by the Boeing–Airbus competition. Internationalization and subsidies are still an important part of the aircraft engine sector, with the British government still providing launch funding to Rolls-Royce.

The Policy Context for Military Aerospace Offsets

Kenneth Flamm
The Brookings Institution

Dr. Flamm presented an overview of the policy issues regarding offsets in military aerospace. Cautioning that his paper was a deliberate attempt to be provocative with the hope of stimulating discussion, he added that he agreed with Dr. Mowery's conclusion that, with respect to the military sector, offsets are not necessarily the most important piece of the story. In fact, there is a much more interesting and complex story.

It is important to begin with an understanding of the cost economics of the military aerospace sector. The industry is driven by economies of scale, particularly because of the large development costs that characterize weapons systems. Fixed production and tooling costs add to economies of scale, as does the dynamics of learning economies. The more you produce, the cheaper the production process because you learn how to do it better. Thus, economies of scale are very important. The higher the volume, the more fixed development costs and production costs can be spread across the entire production run and the greater the learning effects.

As a result, with the worldwide defense downsizing, exports have become critical. This is especially true for non-U.S. producers. U.S. companies still enjoy a large domestic market, with U.S. industry accounting for roughly half of world sales. Everyone else competes for the other half of the market. That puts non-U.S. producers at a distinct disadvantage—basically they need exports to maintain essential economies of scale, or they die.

During the Cold War, production volumes were high enough to maintain economies of scale for almost all countries. Now procurement budgets in European countries have fallen even further than in the United States. This means that the critical mass of volume needed for a new weapons system must rely on export sales.

In the United States, exports continue to play an important role in lowering the cost of weapons systems. However, the U.S. volume of R&D and procurement is large enough to maintain a strong defense industry without exports. This gives the United States the luxury of contemplating policy without having to deal with the immediate short-term requirement of exporting to maintain production. The Europeans, on the other hand, find themselves in a very different situation, with a driving need to export.

One consequence of this need to export is a tendency by our competitors to export more-advanced capabilities as a way of competing with U.S. industry. The result is a vicious cycle. The response to this higher-capability competition by U.S. producers is to also export more military capabilities. At the firm level, the logic is straightforward: If the United States does not give the buyers this capability, they will still be able to acquire it from a competing producer. The United States may as well make the sale and gain the economic benefit.

International Cooperation

Complicating matters is the close technological cooperation between the United States and its allies. Historically, the United States has, to some extent, subsidized the development of the very capabilities that are driving the competition in foreign markets. The United States makes up a little less than half of total Allied procurement. However, the United States provides about 72 percent of total R&D. Either the United States is incredibly inefficient in its R&D, or it is subsidizing the rest of the world. During the Cold War, this was a perfectly reasonable policy for building up the defense industrial base of U.S. allies. From the point of view of economics, it also makes sense to sell off components or technology to our allies to recoup some of the money already spent to develop the technology.

All of this strengthens alliance ties by creating a world of close cooperation. The subsidies to Allied defense industries have been both direct (through licensed production, coproduction, technology licensing, and participation by U.S. companies in product development teams in Europe) and indirect by providing components for European weapons systems so that they do not need to develop a particular technology embodied in the component themselves. The United States has thus created a system in which it cooperates technologically while competing for sales.

With Negative Consequences

The above is a consequence of a logical set of actions by the individual actors that has significant negative impacts for the system. The intensified competition for sales has led to increased transfer of higher levels of technology. This has a negative feedback on the United States in terms of increasing the technological capability of foreign threats. As a result, it becomes more important for the United States to maintain its technological lead by building even more-advanced weapons systems domestically. Thus, the United States has a system whereby it cooperates technologically to create its own competitors. The intensified competition increases the transfer of advanced capabilities overseas, resulting in a heightened threat that then requires higher levels of domestic R&D to maintain a technological lead.

There are two possible scenarios for the outcome of this dilemma: a much higher level of defense spending or a more unsafe world for the United States. The preferable third alternative is to find some way of creating a new system with our allies. Such a system would preserve the desirable elements of cooperation but moderate the unhealthy competition to make sales to questionable customers and markets.

Dr. Flamm noted that he is confining his analysis to military sales. These sales, however, constitute 50–60 percent of total aerospace industry sales—even though this year may constitute a turning point in which commercial sales are greater. Even in the aerospace engine sector, military R&D continues to play a very important role in developing new jet engines. Thus, military sales continue to be a major part of the aerospace industry.

Policy Issues: Export Licensing and Advocacy

Turning to policy issues, Dr. Flamm outlined three important ways in which the U.S. government supports foreign exports of defense systems—which do not include offsets as they are not a critical factor. First, export licensing is where the government has the most leverage over export sales. As explored in greater detail in the paper, the export licensing system remains a hodgepodge of bureaucratic warfare in need of reform. Export licensing is key for military exports. Number two on the list is diplomatic and administration support for exports. Government officials spend a great deal of time arguing on behalf of U.S. weapons systems with foreign buyers. Although this form of government marketing support is not talked about a lot, it is a more important factor in military sales than many other forms of government support.

. . . And Financial Subsidies

The third form of government support is financial subsidics. Ironically, the way in which some of the subsidies are structured tends to defeat the purpose of exports sales. For example, R&D recoupment costs are routinely waived for exports. But the purpose of exports from a defense-industrial perspective is to maintain economies of scale to overcome high development costs. Waiving the recoupment costs means that you are selling to foreign buyers at marginal cost and not charging them for the costs of development. Any economic return to the DoD from spreading these development costs over a larger volume of sales is immediately lost. But the United States does this because its competitors do it. There is an additional, similar benefit to foreign companies when the United States sells them components without charging R&D costs, which they then use in selling their own weapons system in competition with a U.S. weapons system.

Another issue, in terms of financial subsidies, is loan guarantees for military sales. As is argued further in the paper, these broad-based subsidies for exports do not seem to do much to generate U.S. overseas sales.

Offsets are very far down on the list of ways in which governments support exports. The U.S. government has no official policy on offsets. Each part of the government has its own view. Unofficially, the policy of the DoD seems to be that the United States will grant whatever offsets it takes to win the deal.

Cooperative Programs to Transfer Technology

Technology transfer through U.S. cooperative programs, licensed production, coproduction, and codevelopment programs are significantly more important than most offsets in diffusing technology. Citing a National Academy of Sciences' report on Japan[2], Dr. Flamm disagreed with Dr. Mowery, in that he believes there are a number of examples in which military technology transfer has had an impact on commercial aerospace technology. Western Europe started out with a high commercial aviation base and therefore was not as strongly impacted by military technology transfers. But it is hard not to conclude that military programs have had a positive impact on the Japanese aerospace industry.

Two Fundamental Issues

In conclusion, there are two fundamental issues concerning offsets. The first is technology transfer through offsets. If the technology being transferred was funded by the company, then it is a purely private transaction with the company best able to make the decision. However, in the case of government-funded technology, the government must play a broader role as the custodian of the public interest. It is quite possible for a company to make a rational decision on the transfer of technology that might be detrimental to the entire industry, especially when the technology transfer impacts the competitive position of other parts of the industry. Thus, there is a serious potential mismatch between private and public interests when considering technology transfer of government-funded technology.

---

[2]National Research Council, *High-Stakes Aviation: U.S.–Japan Technology Linkages in Transport Aircraft* (National Academy Press, Washington, D.C., 1994).

The other fundamental issue concerning offsets is the impact on the trading regime. For example, dumping is a way of life in the defense industry. No one ever sells for the fully loaded average cost of production of a system. Systems are sold at marginal cost; thus, in this sense, everybody dumps. Given this, aerospace has a trade regime that is different from all other goods. Moving to a more open trading system will require reinventing the rules for aerospace. Industry often complains that it wants a level playing field. There are two ways to get a level playing field: Either reach agreement with the major exporters in the area or use countervailing duties. Countervailing would lead to a competitive spiral, because European governments and European industry cannot simply give up attempting to maintain their own domestic production. Therefore, a more complicated solution for the trading regime is needed.

DISCUSSANTS

Page Hoeper
U.S. Department of Defense

Mr. Hoeper stressed his agreement with an earlier speaker, Mr. Beckman, as to the difference between economics and policy. It is important to note that economics is a tool for measuring the impact of policy; it is not policy. He also pointed out that underlining the discussion is the distinction between a free market economy and a planned economy. Globalization is essentially the peace dividend—the thing that has given us economic growth, productivity growth, and low inflation even as it has resulted in unwelcome income disparities. As a result, the United States is always looking over its shoulder at how other countries are operating. Specifically, questions continue to be raised as to whether more-directed economies, such as Japan and China, will do better. Even with the end of the Cold War and the current Asian meltdown, these countries are not yet willing to change.

Mr. Hoeper stressed that the role of the DoD is national security. The DoD view of how military operations will be conducted in the future, as outlined in Vision 2010, is one of joint operations among the services and coalition operations with U.S. allies. Given the importance of information technologies, coalition operations will require interoperability and common systems. The key question from the security point of view is how does the United States get those common systems without economic distortions?

Addressing the Prisoners' Dilemma

In the past, the United States had a unifying threat that encouraged our allies to buy our systems and not to sell to bad actors around world. Now, we are in a situation of intense competition in which each national industry needs to be supported by foreign sales. This leads to what economists call a "prisoners' dilemma" in which one of our competitors would sell an advance system to a country to whom we would not normally sell. But, because our competitors are going to do it, we feel forced to do it as well. The result is a downward spiral of increased proliferation of more-capable weapons at lower and lower prices.

The only solution to a prisoners' dilemma is cooperation or collusion. Dr. Flamm, in his paper, makes reference to the need for some form of "inner-circle" approach to solve this problem. Mr. Hoeper expressed his concern that this might result in a decline in competition. However, it is important to have some mechanism of cooperation to get out of this prisoners' dilemma without driving values to the bottom and continuing to proliferate weapons. A solution for future weapons systems may be to develop competing transatlantic teams in which we agreed in advance on the military needs for the alliance and on the percentage of the production that the United States and Europe will each buy. The United States and Europe could then split the work, including development, in order to capture the economies of scale.

This only works for future systems. For systems that are already developed, there is already a rush to the bottom. But it is time to compete at least two transatlantic consortia to satisfy the needs for coalition operations in the future.

Frank Parker
ITT Defense and Electronics

Escalating Requirements

Mr. Parker noted that the industry is very concerned about the issue of offsets. The demand for offsets has increased rapidly over the past few years to a point where approximately 110 countries now impose offset requirements. Fifteen years ago, the number would have been 20 countries, with much lower levels of offset demands. Given the range of the various types of offset requirements used by these countries, it is very difficult to define what these offset programs are. Although the focus is often on prime contractors, offsets also have a considerable impact on the sub-tier. A pass-through to a subcontractor imposes a tremendous burden, especially if that subcontractor does not have the infrastructure in place to meet that offset requirement.

The issue of technological transfer, as addressed in Dr. Flamm's paper, is of special interest. No company wants to give away its technology and create another competitor. However, there is great strength within the U.S. industrial base, and companies have developed a number of tactics to meet offset requirements. In each case, companies must devise a unique way of how best to handle the particular requirement depending on the country.

Technology transfer may not be as serious a problem as outlined in Dr. Flamm's paper. Companies do everything they possibly can to mitigate the amount of technology that is transferred. In contrast to Dr. Flamm's comment, Mr. Parker believes that the U.S. State Department does a reasonably good job in trying to curb some of these technology transfers in its export licensing process. The problem is lessened because of the fact that most of the products sold internationally are mature, sometimes to the point that they may almost be obsolete.

Mr. Parker also disagreed with Dr. Flamm's comment that there is no official offsets policy. The Defense Production Act of 1992 originated from a 1990 presidential statement of policy. The law states that the policy of the U.S. government is to not interfere in the issues of offsets. This is a position that industry supports. Industry recognizes that offsets can be an important competitive weapon. However, Mr. Parker suggested that it would be appropriate to restrict the use of government money to support offsets in the foreign military sales (FMS) program.

Mr. Parker closed by stating his support for the broader argument in Dr. Flamm's paper that there are more overriding issues affecting the industry than offsets.

GENERAL DISCUSSION

Albert Kelley, Massachusetts Institute of Technology: Dr. Kelley stressed the importance of separating the military and commercial sectors of aerospace industry, as was done in Dr. Flamm's paper. As was pointed out earlier, venture capital is especially important in commercial aerospace, whereas government procurement is key to military aerospace. Military aerospace is a less risky business, but more subject to government regulation. Regarding one portion of Dr. Flamm's paper, Dr. Kelley asked whether the issues raised in the paper concerned the FMS program or concerned offsets. Many of the issues are raised as part of a FMS package. But such packages often include offsets, thereby confusing the analysis.

Kenneth Flamm, The Brookings Institution: Dr. Flamm replied that he agreed with Mr. Hoeper's remarks concerning the need for a new way to jointly develop future weapons systems with U.S. allies. Such a transatlantic or transatlantic–transpacific mechanism could be constructed in a way so as to increase competition. His suggestion in the paper for an "inner circle" was one form this new mechanism might take. Creating such an open market in defense procurement with U.S. allies would create a larger volume. Enabling easier construction of multinational teams to compete on some systems is an appealing idea. This would ease the worry of our European allies about the disappearance of their industrial base while at the same time encouraging competition.

Dr. Flamm disagreed with Mr. Parker's comment that the State Department does a good job on export licensing, or even has a good understanding of technology issues in general. He reiterated his belief that export licensing is a process in need of improvement, particularly in terms of its coherence and in the need to articulate broad objectives for technology transfer policy.

Technology transfer is clearly an area for potential mismatches between public and private incentives. Specifically, technology largely funded by the government can be part of private deals, which may not take into account some of the objectives of the government when it funds the development of technology. He also stressed that the defense industry can in no way be considered a truly free market—DoD has direct control through a number of ways over the configuration of the industry.

Dr. Flamm agreed with Dr. Kelley's comment that, in discussing FMS, he barely mentioned offsets. Dr. Flamm stressed that the reason for this was his belief that offsets are not the most important dimension of the issue. In addition, FMS is less and less important in military sales.

John Sandford, Rolls-Royce, N.A.: Mr. Sandford reminded participants that FMS financing deals take much more than a year, even though there is an example of Lockheed putting together a deal very quickly.

Joel Johnson, Aerospace Industries Association: Mr. Johnson pointed out that, whereas export licensing is an important government export policy instrument, it is a case of easing a negative barrier on the industry rather than actively helping. In addition to the policy concerned, there is also an administrative problem with export licensing. In the commercial business, it is required to get parts to customers within 48 hours. Export licenses can take from 11 to 90 days. This is a competitive disadvantage for U.S. exporters.

Concerning the issue of recoupment, Mr. Johnson argued that, by the time a product is being exported, it is old enough that it is no longer possible to load development costs into the price. In a commercial business, sunk costs are loaded up front when you think you have market advantage; you do not average the costs over the life of the product. The DoD insists on trying to do that, and industry only asked for common sense. The system of collecting R&D recoupment made no sense in the commercial world, and the DoD never benefited because the funds went directly back into general revenue.

General Discussion of Dumping: A number of participants commented on Dr. Flamm's statement about dumping. Gordon Healey stated that aerospace companies are not aware that they are dumping; companies do not remain in business by failing to make a profit. Dr. Flamm responded that failing to make a profit was not the definition of dumping. Dumping is selling below the fully loaded average cost of production. The term selling below "fair market value" was also used in the discussion. Mr. Beckman noted that the issue raised an important question as to what is the market—there is not a free market in defense products but one where there is only one buyer. Dr. Flamm commented that there are mechanisms for calculating dumping where markets are thin. Mr. Johnson suggested that the issue was more one of a countervailing subsidy case under U.S. trade law rather than dumping. Companies are not selling below their costs. The issue is how to deal with government subsidies. Seventy percent of sales are sold government to government with the price constructed by the government. The revenue to the companies would be the same whether or not there was R&D recoupment. The difference between waiving R&D recoupment and not is the difference between making the sale or not. Thus, the issue is not a defense industry issue, but a government policy issue.

Dr. Flamm agreed with this last observation, noting that the important point is that the normal practice in FMS is to sell below the average cost being charged to the U.S. government—the average product-cycle costs of the system. Mr. Healey disagreed with Dr. Flamm's last point. Dr. Flamm reiterated that he is not criticizing the industry, merely pointing out the rules of the game over which industry has no control. Dr. Wessner noted that the discussion on dumping is recurrent and controversial. It was explored in some depth by a previous National Academy of Sciences' study on trade in technology, although no consensus was reached on the issue. The STEP report, *Conflict and Cooperation in National Competition for High-Technology Industry,* summarizes the competing views and provides useful source material.[3] A companion volume, *International Friction and Cooperation in High-Technology Development and Trade* has two relevant papers. The paper by Thomas Howell is especially relevant. It describes the dumping of steel products at the turn of the century by producers in the United States and Imperial Germany (and the subsequent security consequences for the United Kingdom during World War I) and draws parallels with the more recent cases of dumping in semiconductors.[4]

John Shaw, Cambridge Consulting Group: Mr. Shaw agreed with Dr. Wessner's emphasis on the need for historical perspective. He suggested that much of the discussion assumes a continuum on the issue of military export sales. He stressed, however, that over the past 20 years there have been a number of different preoccupations, such as "merchants of death," foreign corrupt practices, and the end of the Cold War. Much of the moral posturing on the issue has subsided, but the overall problem remains: There is no overall policy. Policy keeps changing and various parts of the government get involved on an ad hoc basis. He noted that the current preoccupation is national economic security, with the central issue being the coordination of government activities with various private sector initiatives. This preoccupation has taken central stage in part because of the lack of other issues. Mr. Shaw cautioned that, based on the history of this topic, the industry is likely to be challenged by some new concern in the future.

---

[3]See pp. 80–85 and Box G, "The Dumping/Antidumping Policy Debate" in *Conflict and Cooperation in National Competition for High-Technology Industry* (National Academy Press, Washington, D.C., 1996).

[4]See T. Howell, "Dumping: Still a Problem in International Trade," in National Research Council *International Friction and Cooperation in High-Technology Development and Trade* (National Academy Press, Washington, D.C., 1997).

Panel 3

The Effect of Offsets, Outsourcing, and Foreign Competition on Output and Employment in the U.S. Aerospace Industry

Robert E. Scott
Economic Policy Institute

Dr. Scott stated that, although his paper looks at the employment effects of offsets and other types of trade, his comments will focus on issues raised earlier during the symposium and on policy options for dealing with offsets and other factors affecting employment in the commercial sector.

He began by noting that the future of the industry may differ significantly from its past, particularly regarding employment. Over the past seven to eight years, the industry has gone through a massive downsizing driven by declines in defense expenditure. Offsets were a relatively small contributor to employment loss in the past, as pointed out by earlier speakers. Dr. Scott suggested, however, that as downsizing approaches its conclusion, trade in general, and offsets in particular, are likely to be bigger factors in employment loss in the future.

There has been a lack of concern by many analysts over the impact of trade on job loss in the aerospace industry. In part, this is based on the general presumption within the economics profession that over the long run trade will be balanced. The assumption is that if we do not sell aircraft, we will sell something else to the rest of the world. However, the U.S. has been running a sustained trade deficit for almost 20 years. In light of the developments in East Asia, trade deficits are likely to increase dramatically in the future, not decrease. One consequence of continuing trade deficits and increased competition, particularly with developing countries, is increasing downward pressure on wages. The loss of jobs in a high-wage sector such as aerospace is of particular concern, especially for labor unions.

Dr. Scott took exception to earlier statements that globalization has generated sustained levels of growth and higher levels of income. Production workers, who make up three-fourths of the labor force, have seen their real incomes fall substantially over the past 20 years. Income inequality has increased between production and nonproduction workers. Most economists believe that trade is responsible for at least 20–25 percent of this increasing gap between production and nonproduction workers' incomes. Thus, there is a significant question in the minds of production workers, who constitute the bulk of the work force and the electorate, as to the benefits of globalization. This is the reason for the growing public opposition to globalization, as expressed in the fast-track debate last fall and current debate over the International Monetary Fund (IMF).

New Developments

Dr. Scott pointed out that it is also important to look at how the Asian crisis will affect the economics and the politics of aerospace industry. There is some analysis that suggests that the U.S. trade deficit will grow from $200 billion today to $300 billion within the next 18–24 months. Dr. Scott's preliminary analysis is that this $100 billion increase in the trade deficit will result in the loss of approximately 1 million jobs, mainly concentrated in manufacturing. The total unemployment may not rise if the Federal Reserve lowers interest rates, but massive sectoral shifts in employment could still occur. The transportation sector, for example, is likely to lose 5–6 percent of current employment.

The result will be a very different political environment. Already there has been a large number of cancellations of commercial aircraft orders. This will create something of a crisis within the aerospace industry, which may present an opportunity for some of the changes talked about during this symposium.

Employment Impacts

Turning specifically to the employment impacts of offsets, Dr. Scott pointed out that employment in aerospace peaked in 1989, as shown in Table 1 of his paper. Since that time employment has dropped from approximately 1.3 million workers to a low in 1995 of about 780,000 workers, before recovering in the past few years. Employment in civil aircraft followed the same general trend. The paper dissects that employment decline into three causes:

- a decline in sales, dominated by the declines in defense, but also in the commercial sector, which is responsible for approximately one-half of the job loss;
- outsourcing, meaning increasing imports of parts and components, which accounts for 6–10 percent of the job loss; and
- rapid productivity growth, which accounts for the remainder of the job loss.

Thus, Dr. Scott agrees with the earlier analysis by Dr. Mowery that in the past offsets played a relatively small role in explaining the decline in employment.

However, the coming two decades will see a sharply different environment. Defense spending is going to be, at best, constant. The commercial sector will grow in importance. Table 6 in his paper presents a revised version of the analysis of the future impact on unemployment of offsets presented earlier in *Jobs on the Wing*.[5] The analysis breaks the causes of job loss into two factors. The first factor is offsets—where the effect is the loss of about 45,000 direct jobs by the year 2013. The second factor affecting employment in the industry is the continued loss of market share in the commercial sector to Airbus. Extrapolating the trend in the decline of Boeing's market share over the past decade or so, the projection is for a loss of approximately 77,000 direct jobs by 2013. Thus, the decline due to market share loss is twice as large as that due to offsets.

The total job loss in the aerospace industry, therefore, is approximately 123,000. Indirect job loss in supplier sectors such as steel and rubber brings the total loss to over 200,000 jobs. This is a drop in employment in the total aerospace industry of roughly 15–20 percent. Because the losses will be concentrated in the civil aircraft sector with a current job base of about 300,000, this will have a significant impact on employment.

Employment decline will also have a significant political impact. Current concern over globalization has been based on a loss of 2–3 percentage points in relative wages for production workers. Future job losses of this magnitude are likely to attract political attention, especially because the losses are in a high-technology, high-wage industry in which the United States should retain competitive advantage.

Dr. Scott pointed out that he believes that these projections are conservative. For example, Airbus may gain market share even faster than assumed in the analysis, as shown in Figure 3 in his paper.

Given that the industry is at great competitive risk, it is important to craft a coherent policy that goes beyond simply offsets. Although it may be difficult to negotiate a deal on offsets, it is clear that we are in a "prisoners' dilemma" with respect to subsidies. This raises the question as to whether we should consider applying an "inner-circle" approach to the commercial sector, as was suggested earlier for the military sector. This would mean looking for leverage points coming out of the crisis in the industry that will emerge in the next few years to negotiate a market share agreement with the European Union.

Notwithstanding the negotiating difficulties referred to earlier by Dr. Robyn, Dr. Scott reiterated his recommendation from his paper for some form of bilateral agreement with the European Union to eliminate the use of offsets as a marketing tool, possibly as an extension of the Foreign Corrupt Practices Act. Dr. Scott concluded with the recommendation for continued dialog among industry, government, and labor as exemplified by this symposium, to build support for the development of a policy dealing with both offsets and the larger competitive threats to the aerospace industry.

---

[5]R. Barber, and R. E. Scott, *Jobs on the Wing: Trading Away the Future of the U.S. Aerospace Industry* (Economic Policy Institute, Washington, D.C., 1995).

DISCUSSANTS

Howard Rosen
Joint Economic Committee

Dr. Rosen began by stating that he agreed with the premise in Dr. Scott's paper as to the severity of the problem and the devastation caused by the economic dislocation. His comments, however, focused on two broader and possibly provocative points. The first is the fact that the economy has been going through significant structural changes in the past 25 years. It is therefore important to look at longer-term trends. The U.S. economy today is very different than it was 25 years ago—and changes in the industry need to be seen in that context.

The second point concerns the importance of trying to quantify economic impacts. As someone who was previously involved in building models of employment effects, Dr. Rosen commended Dr. Scott's efforts at analysis. Especially important is the use of the analysis to understand the magnitude of the issue. However, he reminded participants that these numbers are the beginning of the discussion, not the end. He cautioned against crafting policy using numbers that may only be strong enough to disaggregate the causes of job loss and help understand what happened in the past, not necessarily predict what will happen in the future. He also expressed concern over the specificity of the numbers and suggested using percentages instead.

Dr. Rosen went on to note that the trade debate over the past decade has tended to focus on two separate issues: free trade versus protection, and how to trade, for example, fair trade. Now comes the Asian crisis, which no one saw coming, that overwhelms all other issues. This shows that for too long we have been ignoring major changes in exchange rates and other factors, which have the possibility of overshadowing all other effects.

Employment Policy, Not Offsets

Seeking to broaden the discussion, Dr. Rosen suggested that the central issue is not offsets, but employment policy. He raised the question as to whether the United States is using its trade policy to create an employment policy. Looking back over the past few years, there seem to be two forces at work on the industry: the end of the Cold War and a dramatic increase in competition. Both of these are positive developments, even if some of their consequences are negative. Thus, it is important to think the issue through clearly.

Dr. Rosen further commented that, although he appreciates the concern over the dislocation within the economy, he is concerned with the connotation that all unemployment is bad. As Dr. Scott's analysis shows, the employment effect of productivity gains—which is generally viewed as an economic positive—is double that of offsets. This suggests that the topic is more complicated than just the employment effects.

He then raised the question of the importance of the aircraft industry. The answer, he suggested, is that the industry is big; however, the size can cause a distortion in the data. Other industries are being hit proportionately much harder. But because of the size of the aircraft industry, the aggregate numbers look much larger.

Thus, the question must be asked as to what is the goal of policy: save an industry, regulate trade, or create an employment policy to improve living standards for all Americans? If the goal is raising living standards, it is not clear that focusing on offsets tells us how to do this. Interest rates, for example, have a larger impact on unemployment in the industry than do offsets. Policy makers, businessmen, and labor leaders need a broader perspective to incorporate these other factors, such as currency and exchange rate policy.

He pointed out the core finding of Dr. Scott's paper that productivity gains contributed to most of the job loss and the declines in sales may reflect a positive development, namely the end of the Cold War.

In addition, Dr. Rosen expressed concern about the focus in the paper on gross employment changes rather than net employment changes. The analysis should include job creation by exports and attempt to look at whether offsets contributed to those exports. The paper also needs to address the benefits of productivity to the industry.

Dr. Rosen concluded by noting that the most important finding in the paper is that 62 percent of production is in the United States, yet 75 percent of the job loss is also in the United States. This suggests to him that the United States is absorbing more of the job loss than other countries. U.S. employment policy is to put adjustment costs on the backs of individuals. This calls for a serious policy response, similar to a pilot program that is being conducted in New Mexico to help coordinate economic development and assistance programs to meet specific local needs. Dr. Rosen complimented Dr. Robyn in her role in helping to bring about this innovative pilot program.

Gordon Healey
Defense Industry Offsets Association and Bell Helicopter

Mr. Healey began by describing the Defense Industry Offset Association as an organization of 65 aerospace prime contractors who meet regularly to discuss and exchange information on offsets. He stressed that, in the defense industry, there is no such thing as a voluntary offset. There are offset agreements that may not be dictated by law or government regulation, but that are dictated by the competitive environment. Offsets are an inevitable part of marketing defense products overseas. Companies only have the option of either engaging in offsets or walking away from the deal.

Concerning the employment impacts of offsets, Mr. Healey stressed the importance of offsets in maintaining jobs. Total sales of the industry are approximately $130 billion—$80 billion domestic, $50 billion international. If the industry chose not to do offsets, that $50 billion in international sales would disappear. Those $50 billion in sales translates into approximately 300,000 jobs. As an element of the international marketing process, offsets are one of the greatest factors contributing to aerospace employment.

Disagreeing with Dr. Scott's conclusion that offsets are an important threat to domestic production, Mr. Healey affirmed that offsets are in fact an important key to domestic production. As an inevitable part of the international sales process, offsets are a contributor to those 300,000 jobs.

At this point, a question was raised by Brad Botwin, U.S. Commerce Department, concerning the data. Industry data supplied to the government show $3 billion in offsets, not the $1.5 billion shown on Mr. Healey's chart.

Mr. Healey responded that in 1997 the industry did engage in approximately $3 billion in total offset agreements. Half of that, about $1.5 billion, were direct offsets in the aerospace sector. That translates into 10,000 jobs directly affected by offsets.

Mr. Healey went on to stress that offset managers try very hard to mitigate the effects of offset requirements. Companies have many ways to minimize the impact of offsets on their work force and supplier base. These include

- negotiating down the percentage of the offset required;
- extending as far forward as possible the period of performance for the offset,
- obtaining the most favorable mechanism for offset credits in the contracts such as high multipliers, and
- ensuring the longest possible list of parties eligible to perform the offset.

As noted above, multipliers inflate the actual value of offsets. For example, multipliers are offered as incentives by customer countries to place offset projects in certain areas. The value of the multiplier credit can be from two to ten times greater than the actual value of the work performed.

Another mechanism to satisfy offset requirements includes sending personnel to the customer country to provide technical assistance and training. Such assistance is also often associated with multipliers. In addition, offset credits can be gained by offering assistance with financing and in marketing customer-country products in the United States.

Joint ventures and investments in customer countries is another way of gaining offset credits, and they often show up as indirect offsets. In some cases, these arrangements can be very beneficial to other U.S. firms. Mr. Healey gave the example of one joint venture with a small U.S. firm with environmental technologies that helped that company expand into foreign markets. The defense contractor received offset credits for the amount of business done by that joint venture and for the investment made by the joint venture in the customer country. This form of indirect offsets is becoming more common.

Policy Recommendations

Mr. Healey concluded with a discussion of policy recommendations. He stressed the industry's belief that the United States must refrain from taking unilateral action on offsets. This would be tantamount to walking away from international sales. Second, the data collection process needs to be improved. Industry representatives have met with Commerce Department officials and have made suggestions on how to make these improvements. Third, multilateral discussions on the issue should continue. It is up to our trading partners to help restrain the use of offsets. Fourth, the dialog within the United States must continue so that all interested parties better understand one another.

Finally, Mr. Healey reiterated the findings of earlier speakers that offsets are not the largest problem facing the industry, nor are offsets exporting large numbers of jobs. In today's competitive global market, companies simply cannot walk away from offset obligations. If there are problems with offsets, they should be fixed. But he urged that the issue be kept in context.

GENERAL DISCUSSION

Page Hoeper, U.S. Department of Defense: Mr. Hoeper suggested that the demise of the defense industry may be overstated. He quoted from George Bernard Shaw's *Major Barbara* where the concern over the business impact of peace is answered when the armaments officer states, "Fear not, I have faith in human nature."

General Discussion on Data: Mr. Shaw noted that the $1.5 billion used in Mr. Healey's presentation does not take into account offsets work in the pipeline, which would raise the figure considerably. Mr. Healey replied that this was simply offset performance in one year, and solely in the aerospace industry. Mr. Botwin observed that the $130 billion in Mr. Healey's presentation is total aerospace sales. Commerce Department's figures (and Mr. Healey's $1.5 billion figure) cover only defense offsets, with no data as to the size of offsets in the commercial sector. Using the Commerce Department's figures and excluding those defense sales with no offsets, offsets make up 80 percent of the value of the sales. A number of participants pointed out that this figure is for new agreements announced in 1997.

Mr. Healey argued that this does not mean that 80 percent of the production is shipped overseas. The industry works hard to minimize the impact of offsets through a variety of ways. Companies try to create for the customer what to them has a certain level of value, but which actually costs the company only a fraction of that number. A number of participants questioned whether the 80 percent figure included the multipliers given by foreign governments when calculating offset credits. It was pointed out that $1.5 billion is the value that is actually being transferred due to the offset, which is only a fraction of the offset credit reported in dollars. The 80 percent figure uses the actual amount of offsets credited by the customer.

It was suggested that this discussion illustrates why the data collection effort needs to be improved—and why the industry does not want to have the data released to other customers so that they can compare the multipliers. Data on offsets reported to the Commerce Department are offset credits as measured in dollars. Because of multipliers and other factors, these dollars do not represent the true impact of the offsets in United States, but represent the value to the customer. For example, $1 of production transferred overseas may earn $100 of offset credits from the customer. That does not mean that the cost to the United States was $100; the cost was only $1. In 1997, $15 billion in overseas defense sales was accompanied by $1.5 billion in offset credits; that is the value calculated by the customers of the fulfillment of the offset requirements.

Mr. Healey again affirmed that many people are employed through the offset process—not in doing offsets but benefiting from the international sales that are made possible through this very small amount of offsets.

John Tucker, Commerce Department, attempted one last bid to clarify the numbers. According to data given to the Commerce Department, offset transactions for fulfillment of previous agreements in 1996 were $2.9 billion, whereas the credit value was $3.1 billion. New agreements in 1996 were $3 billion in sales and $2.4 billion in offset agreements, which is about 80 percent. Lawrence Bertino, Boeing Space Systems, pointed out that the obligations for these new agreements are likely to be spread out over 10–15 years.

Better Rules of the Game

Robert Scott, Economic Policy Institute: Dr. Scott responded to earlier comments by stating that his numbers are not very far apart from those of other speakers. His estimate is that less than 10 percent of the job loss is due to offsets. Including commercial production with military production, and looking at U.S. production as opposed to world production, his figures are not that far apart from Mr. Healey's.

The critical issue is whether offsets buy foreign sales. Industry representatives rightfully speak from the point of view of an individual company trying to make the sale. However, if the rules of the game could be changed to reduce the incentive to engage in offsets, then all firms would benefit. What is in the interest of any one producer in any individual transaction must be distinguished from how the market is performing. Policy must focus on how to change the rules of the game to improve the performance for both the firms and the country.

John Sandford, Rolls-Royce, N.A.: Mr. Sandford commented that the model could be improved by looking at the impact of airline consolidation. Such consolidation may result in the emergence of buying groups that would have a greater ability to demand offsets. He also stated that the increasing need for access to capital will make it more difficult to distinguish what is an offset. Companies make arrangements with risk-sharing subcontractors to gain access to capital. That capital is needed not only because of the risk, but also because of the need to return value to the shareholder.

In response to a question from Dr. Wessner as to whether all major players in the industry need to satisfy shareholders, Mr. Sandford stated that there was a large imbalance. It is mainly, but not exclusively, the American companies who have to satisfy short-term shareholders. It is his opinion that in today's environment, you could not launch a new engine program without significant risk capital from overseas. That capital would only flow for reasons of increased jobs.

Offsets in the International Marketplace: An Aerospace Industry View

Joel Johnson
Aerospace Industries Association

Mr. Johnson agreed with earlier comments that, from an economic perspective, offsets are not very important relative to other factors in the aerospace industry and the economy. Offsets are, however, politically important, which is the reason for the attention.

International and Domestic Offsets

Because the customers in the defense sector are governments, they want more out of their procurements than simply price and performance; they also want jobs. The industry faces these pressures even domestically in the form of a number of "domestic offsets," such as minority business set-asides and informal requirements for a large geographical distribution of production. The industry is skilled in meeting these domestic offsets demands of its customers, which gives it a competitive advantage when faced with similar requirements overseas.

Buoyant Sales

As noted in Mr. Johnson's paper, sales in the industry have increased over the past year from $112 billion to $130 billion, employment has increased by approximately 40,000–50,000 workers, exports are at an all-time record of $50 billion, and the sector's trade balance hit an all-time record of $34 billion. This is not a picture of an industry in decline.

More important, more than half of annual sales in 1997 were to non-U.S. government buyers—something that has happened only once before in the history the industry, in the 1930s. Of those sales that did not go to the U.S. government, 75 percent were exports. In the future, it is clear that the U.S. government share of sales is not going to go up. The only expansion will be to non-U.S. government buyers, most of which will be international. Given that U.S. government procurement is essentially protected, competition will be concentrated in this growing international market.

Two Trends

The industry has had to cope with two major trends over the past few years. The first has been the decline of the defense and airline markets. The airline market is recovering, but defense spending is unlikely to increase to previous levels. The second trend is the major restructuring within aerospace industry. The industry has dramatically increased productivity as customers have become more cost conscious. Of particular interest was the earlier analysis presented here showing that productivity gains accounted for twice as much employment loss as offsets. Extrapolating these losses forward, the industry will have a negative 100,000 workers by 2013—which makes one suspicious of such projections!

Mr. Johnson has seen little evidence of increased foreign penetration in aerospace parts. Imports over the past few years have held relatively constant at 5.5–6 percent of production. Problems facing the supplier base seem to be more the result of a decline in business and a rationalization of the supplier base rather then increased foreign penetration.

He reminded participants that the United States essentially has a 100 percent offsets requirement. Contracts for almost every major weapons systems require 75–100 percent production in the United States. The question becomes what the United States would be willing to lay on the table in negotiations with its allies on offsets. He suggested that negotiations limiting the United States' ability to demand production here would be politically unpopular.

Concerning the data, Mr. Johnson commented that the three sectors in the Commerce Department report on offsets—gears, machine tools, and shipbuilding—have all been either government subsidized or protected by Section 301 trade actions. In those sectors, offset performance was less than 1 percent. Another concern was that the Commerce Department data did not show the degree to which one foreign supplier was being substituted for another. Given that in these three sectors there is a lot of substitution of one foreign supplier for another to gain offset credits, the 1 percent is probably an exaggeration. Thus, he argued, there is no "smoking gun" as to the negative effects of offsets.

He suggested that a better way to conduct the analysis would be to look at the problems facing the subsectors, rather than start with offsets. Offset managers tend to circle around the subsectors in trouble—those subsectors that are not providing performance, are not cost competitive, do not provide quality, have not taken care of the customer, and have sought government protection in the past. In many cases, the offset is not the sector's problem; the sector has problems that attract offset managers. Offsets may simply be a scapegoat for these underlying structural problems. Anecdotal data about problems caused by offsets may be suspect unless the analysis asks questions about the deeper structural problems.

No Unilateral Action

The industry has a number of recommendations for dealing with offsets. At the top of the list is a plea for no unilateral action. Also important is reform of the export licensing system. Mr. Johnson stated that the single greatest impediment to industry exports remains the U.S. government—because of the problems with export licensing and the tendency to impose unilateral trade sanctions.

He agreed that government funds (through FMS programs) should not be used to support direct offsets. However, there are only two countries that currently use U.S. funds for military procurement: Israel and Egypt. Congress already gives a half billion dollar subsidy to the Israeli defense industry and is therefore unlikely to do anything about offsets in this area. With the FMS program down to $3 billion, FMS-supported offsets are no longer a significant problem.

Negotiations?

The industry remains supportive, but skeptical, of multilateral negotiations on offsets. Mr. Johnson argued that the Europeans would be very happy to agree to no longer offering offsets in third-country markets. The U.S. industry has a competitive advantage in its ability to place work offshore. The Europeans do not necessarily need to make a profit, but find it difficult to place work offshore. If we propose an agreement to limit our ability to place work offshore—and therefore remove a U.S. competitive advantage—the Europeans may well accept it. The situation is different when it comes to transatlantic offsets where there is considerable political pressure on the Europeans to require domestic production. For example, statements from American labor unions concerning protecting American jobs are similar to those made by British labor unions concerning protecting British jobs. Mr. Johnson urged that the issue be kept in context.

He also expressed concern as to the assumption that indirect offsets are worse than direct offsets. The more one moves toward indirect offsets, the closer one comes to real trade. Direct offsets require bilateral sectoral and country balance; indirect offsets move toward only requiring bilateral country balance. The more the offsets are spread around the economy, the less damage is done to the aerospace sector and the more the transaction looks like real trade, because for every import there has to be an export in the economy sooner or later. The question becomes the extent to which governments, U.S. and foreign, get involved in determining what makes up the imports and exports, as opposed to letting the marketplace decide.

Mr. Johnson closed with the admonition to be careful about data. He stated that there is nothing but downside risk to further publication of the data. The better it looks to the industry's foreign customers, the worse it looks to Congress, and vice versa. Data published by the Commerce Department is used by foreign governments to determine how high of an offset requirement they can impose. He argued for a better way of handling the data so it does not make the problem worse.

DISCUSSANTS

Steve Clemons
Economic Strategy Institute

Mr. Clemons stated that he would focus on the political issues revolving around offsets. The offsets issue has thrived in obscurity and ambiguity. As a result, the industry seems to have a strategy that "doing nothing is best." Mr. Johnson's paper offers prescriptions that would be very tough to implement, while discounting all other possible solutions. There is a general agreement that any unilateral action is probably not possible. There is also a generally held view that offsets are a necessary evil as part of the business of doing defense trade.

However, Mr. Clemons stressed that, because of the mix of national security concerns and tax dollars, offsets are not purely a commercial matter. Offsets meddle in market behavior. They reflect the industrial policy objectives of other countries who are trying to use a sale process to create a high-wage aerospace industry. This calls out for a policy response by the U.S. government.

For too long, industry has played the role of saying "do nothing." It appears that the DoD, starting with the June 1997 workshop by this group, has taken a much more pro-offsets position—to the concern of some in Congress. Because labor unions have brought this issue more to the forefront, there is an increased perception of job dislocation and of a problem that is going to increase. Industry, therefore, must become more of a partner in figuring out what to do about the perceived problem.

The political volatility of the issue will only increase. Mr. Clemons gave the example of companies seeking to work with the national laboratories as part of offsets agreements. This raises politically dangerous questions, including the use of tax dollars and national security research by private companies for parochial deals abroad. Over the next few years, politicians are likely to look more carefully at these issues of the use of tax dollars under the mantra of "corporate welfare."

This is very worrisome to those who believe there is a government role in the development of the technology. Just allowing the system to go forward without any kind of constructive thinking about the involvement of industry, the use of national resources such as the laboratories, and parochial defense sales may be reckless.

Use of indirect offsets is also politically volatile, as shown by the recent case cited by Senator Feingold. Commerce Department data show that indirect offsets are rising. Thus, somewhere, someone in some industry or company will feel that they have been hurt by a subsidized effort to gain a defense sale. The defense industry should be worried about the impacts of indirect offsets because of the political consequences of those impacts outside of the defense sector. There need to be some principles of self-restraint established, which industry seems to be reluctant to do. Without such action, the issue of indirect offsets is likely to attract higher political attention.

Mr. Clemons concluded by reminding participants that this is nonmarket behavior. It is not simply a question of letting markets operate. The situation is parallel to that of the state-versus-state competition for new businesses. States have come up with all types of innovative incentives for companies to locate plants within their borders. An entire industry has grown up around providing these incentives. There is the concern in the political world that this same type of innovation is occurring with respect to offsets, which is not very healthy. That is the perception, and industry must therefore be much more actively involved in helping to craft a solution than it has been thus far.

Randy Barber
Center for Economic Organizing

Mr. Barber commented on the need to be clear on a number of points. The first is the difference between defense offsets and commercial sales. Although perhaps there are no such things as voluntary offsets on the defense side, it is important to keep definitions in mind. Earlier speakers drew a clear distinction between offsets as a government requirement and strategic alliances. Measuring these offsets due to specific government requirements is much easier than obtaining data on commercial strategic alliances. Yet there are clearly decisions that companies make voluntarily from a marketing perspective to enter into certain types of relationships.

Mr. Barber agreed with Mr. Johnson in that U.S. industry developed its ability to offer offsets into a competitive marketing tool. It is clear that parts of the industry, such as Boeing, believe that they have a major competitive advantage because they can offer parts of the production process to the potential customer. Mr. Johnson's suggestion for negotiating a standstill on offsets with Europe is based on the fact that the Europeans are now using offsets as a competitive tool as well.

Referring to Figure 4 of Dr. Scott's paper, Mr. Barber disagreed with Mr. Johnson's data. He argued that imports of engines and parts have increased from 8 to 16 percent from 1988 to 1995 rather than staying flat. Mr. Johnson responded that the import–export data mix engines and aircraft. Rolls-Royce engines are counted as an import; Pratt & Whitney engines end up as part of the aircraft export numbers. Thus, the data are incompatible.

John Sandford, Rolls-Royce, N.A., stated that Rolls-Royce holds roughly 20 percent of the market, but added that company claims as to market share are often suspect. Data on the import or export of engines is often suspect as well because the trade data reflect the full value of the product, whereas much of the production takes place in another country. He cited the example of a popular General Electric engine for which more manufacturing occurs in Europe than in the United States. He also expressed concern that it is unclear how the data account for cases such as Rolls-Royce engines sold to Boeing for aircraft sold outside of the United States. Mr. Johnson added that this means that figures on engine imports will increase as Boeing increases market share and exports more.

Mr. Barber reiterated that part of the problem is a lack of reporting requirements on the commercial side. Beyond the fact that there is a fuzzy line between a true "required" offset, a preemptive offset, a strategic marketing initiative, and a technology licensing agreement, there is no data as to what is happening on the commercial side. We really do not know.

This lack of data has been used by a number of earlier speakers to argue that we cannot prove the impact of offsets. Mr. Barber can understand the industry's argument as to why they have been very leery of the collection, analysis, and distribution of this information. However, it is important to get past arguing over numbers that fundamentally have not been collected. There may be a risk of disseminating the information, but it is likely that the industry's customers share this information among themselves anyway.

Mr. Barber closed with his agreement with Mr. Clemons' comment on the importance of perception. He personally believes that the problems of offsets and globalization are real. But the perception is that companies are voluntarily trading production and technology for sales—an undertaking that a generation ago would have been considered crazy. Although this may have become an acceptable idea within business, it is very unsettling to workers. Workers do not have the mobility of capital and the ability to cope with these changes. This mind-set is a dramatic shift in what one would expect in the behavior of an industry that is so closely linked to national defense and is considered to be a premier example of U.S. manufacturing prowess.

GENERAL DISCUSSION

Greg Martin, Lockheed Martin: Mr. Martin noted that the discussion has moved away from a focus on trade-distorting requirements by governments to attention on voluntary activities that companies might engage in as part of good business practices. He expressed the hope that we were not moving toward legislating what U.S. businesses can do in international business practices.

Randy Barber, Center for Economic Organizing: Mr. Barber responded by noting that many speakers have pointed out that the issue is not one of simply offsets. The issue involves trade policy, employment policy, and capital markets policies, to name a few. To the extent that companies' interests diverge from the interest of their workers or the country, these issues must be addressed. Clearly, the issue is broader than simply government-mandated deals, but includes the various ways in which technology and production are traded for sales.

Mr. Martin responded that this is a greatly expanded definition of offsets.

Steve Clemons, Economic Strategy Institute: Mr. Clemons stated that a parallel can be drawn to the question of international codes of conduct. Two years ago there was a great deal of concern that Congress might move recklessly and legislate on the issue by trying to, for example, tie the tax code to corporate behavior. With respect to policies on offsets, very little has changed in the past ten years. Most have simply hoped that the issue would not receive a lot of attention. Efforts were confined to reporting requirements with the mantra of "take no unilateral action." Although there is a concern about Congress doing something, the corporate community has not initiated anything that might work multilaterally or might respond to the major concerns. Industry seems to be waiting for things to happen while hoping nothing happens and refusing to take the lead to define what might be a positive outcome. Now the situation has come down to labor versus industry, because labor and others believe that offsets are having a negative impact and that they are eroding the American technology base. Industry needs to move away from its passive position and get behind a workable multilateral solution—which would greatly raise the probability of success.

Mr. Martin argued that the industry is working on the problem, but is doing so on its own with its foreign customers. The industry is trying to move away from traditional offsets toward good business practices that do not count credits and look toward long-term business relationships. In these types of relationships, U.S. companies might gain access to markets, whereas customers might gain access to financing and complementarities of technology.

Mr. Clemons noted, and Mr. Martin agreed, that this is being done on an individual company basis. Mr. Clemons stressed, and Mr. Martin again agreed, that individual companies have no incentive to defect in this prisoners' dilemma game because if they do so they are left out. Mr. Clemons then noted the situation therefore is one of "market failure." In such a situation, it may benefit industry to come together as a group with unions and government to develop something that no single company could initiate on its own.

Mr. Martin stated that the industry is willing to do this and share as much information as possible without giving up competitively sensitive data. The industry is also willing to work with others on a multilateral basis. However, there is still concern about legislation that, in the name of restricting offsets, would restrict what the industry believes are good business practices that open markets and share risks.

Joe Evans, General Dynamics: Mr. Evans commented that foreign customers are making offsets much more difficult and expensive to implement, adding to the costs of a program. If U.S. legislation were to restrict the ability of American companies to do what they think is necessary, then the response from industry will simply be to walk away from the business. That, as Dr. Wessner pointed out, would have its own consequences for employment and national security.

Steve Beckman, United Auto Workers: Mr. Beckman noted the importance of the distinction between the military and the commercial sectors. What is being discussed as "sound business practices" are not sound practices by the standard of American economics. Economic theory would predict that the United States should be dominant in the aerospace industry because it has the most efficient production. The problem is that the market does not work that way. It is difficult for business to accept that a different mechanism is needed because the market is not working the way it should. In essence, business is making a necessity into a virtue by saying that these agreements are the only way to get access to markets. The traditional way to gain access to markets is for governments to negotiate access agreements and enforce those agreements through trade mechanisms. If the argument is that trade mechanisms and trade policy do not work, then the discussion needs to be enlarged. The fundamental problem with offsets is the imperfections of the market that are being, in many respects, exacerbated by these "good business practices." Such practices do not move toward a common way of dealing with the problem in a manner beneficial to the U.S. national interest as well as the commercial interest of the companies.

Mr. Martin responded by saying that he was focused on market access in terms of what the customer wants, not in terms of what governments will allow. For the same reason that there is a reluctance on the part of some U.S. companies to buy from a foreign source, there is a reluctance on the part of European companies to buy from American sources. This reluctance stems from that fact that suppliers are not located in-country. You have to be there to support your customer. These strategic alliance arrangements allow U.S. producers to be there without having to open up a new company in each customer country.

Charles Wessner, National Research Council: In closing this part of the discussion, Dr. Wessner noted that Mr. Beckman's point appeared to be not that there are market imperfections through lack of information, but rather that the issue is one of how to respond to government interventions designed to achieve a particular goal.

Dual-Use Supplier Management and Strategic International Sourcing in Aircraft Manufacturing

Todd A. Watkins
Lehigh University

Dr. Watkins presented a case study on how one company deals with supplier management and offsets. The case will be presented as the Generic Aircraft Manufacturing Company (GAMC) because of the sensitivity of the data and issues. Originally, the case was a study of dual-use manufacturing. Data collected on strategic international sourcing and offsets, however, proved to be both very interesting and extremely sensitive. As a result, it took two years for the company to agree to release the case study, even then only in generic form.

An Overview

The purpose of the study was to extend the analysis of lean manufacturing practices to supplier management in the aerospace industry. American aerospace companies began moving to the lean manufacturing model in the 1990s as the industry sought to rationalize the supplier base because of the economic downturn. The complexity of the products in aerospace implies that the industry is managing a supplier base at least as complex as that of the auto industry, where the lean model was developed. Mid-tier companies need to deal with the supplier management practices of the upper-tier manufacturers, as well as manage an equally complex series of relationships with their own suppliers. Unique to the aerospace industry, however, is the need to manage a supplier base in the two separate sectors of defense and commercial products. While supplier management practices at GAMC differed little between the two sectors, the major exception was offsets.

The GAMC Profile

GAMC is a producer of major structural subsections, such as wings, for the prime airframe manufacturers. The company supplies subsections for both defense and civil aircraft using a single supplier management system across both sectors. The experience of the company reflects, at a microlevel, many of the issues discussed earlier at a macrolevel—including the mind-set of "if not us, then someone else will get the business" and a lack of data. However, their strategic outsourcing initiatives have come about due to pressure from commercial, not defense, customers.

The company also illustrates the turmoil in the industry, having reorganized many times under various new owners. Both employment and the number of suppliers declined between 1991 and 1995 as the company went from a size of $1 billion to $650 million before being absorbed by a larger firm. The company's mix of business has shifted from 60 percent defense to 40 percent defense.

Mid-Tier Squeeze

The company has undertaken a whole range of supplier management actions normally associated with lean manufacturing, such as collaborating closely with their suppliers, building long-term relationships, and assisting in improving suppliers. Applying these principles while internationalizing the supplier base leads to a situation for a mid-tier company in which the company ends up essentially working to help improve firms that are and will be their competitors. Dr. Watkins labeled this situation "mid-tier squeeze."

GAMC has become more attractive to its customers because it is a leader in applying these principles. GAMC is moving away from a loosely organized supplier system to a much more closely controlled structure, with a number of long-term partnerships with both customers and suppliers. Because more than 80 percent of their suppliers service both defense and commercial products, the company uses one single supplier management and quality control system across both sides of the business.

A clear goal of their supplier management system was to increase the number of foreign suppliers. All of the pressure for this internationalization of the supplier base was coming from commercial customers. By contract, GAMC was required to increase the number of foreign suppliers to a certain percentage. The reason for this pressure was that the company was supplying subsections to aircraft targeted to Pacific Rim markets.

Beginning in the mid-1980s, under severe pressure from their customers, the company went searching for foreign suppliers. Their first strategic outsourcing arrangement was with what Dr. Watkins labels "Nagoya Aerospace," a consortium of five Japanese companies. GAMC transferred to Nagoya the production of a particular substructure that they had been doing in-house for a number of years.

This was a mature, middle-technology part, comprising about 15–20 percent by cost of the total delivered structure that GAMC was sending to the prime contractor for their aircraft, "Norton Alpha." Nagoya was chosen for three factors:

- the bid was even lower than in-house production at the time (this is no longer true due to currency fluctuations),
- the skills of the company for long-term partnering (i.e., will Nagoya be able to do more than build-to-print in the long term), and
- the interest in the company in taking on a mid-technology product.

A year later the product was second sourced to a Korean company. Subsequently, GAMC has made similar arrangements for many other products in many other countries.

Improving Foreign Suppliers

Under the lean manufacturing model of building collaborative relationships with suppliers, GAMC has the responsibility passed down from the top tiers for managing and improving the quality of the lower tiers. GAMC retains the responsibility for the quality of the product. As a result, GAMC spent a lot of time shifting the production of that part over to its foreign supplier. They provided significant technical and managerial assistance. In essence, GAMC was forced to take on a new role of improving foreign suppliers because of pressure from their customers.

Ultimately, Nagoya was given the responsibility for managing the production of that substructure. This means that Nagoya ended up managing the relationship with those U.S. suppliers who had been supplying GAMC for that substructure.

Dr. Watkins pointed out that this form of production sharing makes data collection very difficult. The product coming back to GAMC from Nagoya and the Korean manufacturer contains a significant amount of U.S. components and subassemblies. It is unclear how this gets counted as an offset credit when it goes to the prime contractor, "Norton."

Because this subsection is a mature product, the U.S. subsuppliers are already in place. The question has to be raised as to whether Nagoya will use the same U.S. suppliers for the next product. As it turned out, in one case Nagoya completely by-passed GAMC and went directly to work for Norton. Over time, the relationship between GAMC and Nagoya has expanded. GAMC outsourced a similar substructure to Nagoya in the 1990s.

Significant Payback

GAMC has received a significant payback from this process. They spent a significant amount of time, money, and effort over a decade to increase their ability to work with international suppliers. This international exposure has made them more attractive to other customers and thus boosted their edge in the highly competitive middle-tier supplier market. A new major U.S. customer, Kramden, came to GAMC to work on a new transpacific aircraft. Kramden is attracted to GAMC specifically because Kramden's target market was the Pacific Rim, and GAMC has established relationships with Pacific Rim suppliers such as Nagoya. As a result of the importance of this relationship, Nagoya has moved from essentially a build-to-print shop to being GAMC's development partner on this major substructure.

Policy Issues

Employment

Dr. Watkins then turned to a number of policy issues. In this case, the impact of strategic sourcing on employment was marginal. GAMC went from more than 11,000 employees to about 5,000 employees between 1989 and 1995, which was mostly due to cancellations in defense, not to strategic sourcing. By 1995 the net effect of strategic sourcing was probably positive because of the new deal with Kramden. The longer-run impact on jobs is still unclear. Nagoya may become a viable competitor because GAMC has transitioned its skills to them and essentially introduced them to Norton. This transitioning may have a much greater impact on jobs than the outsourcing, which was probably only about 5 percent of the employment loss.

Dr. Wessner asked why, then, was the case so sensitive. Dr. Watkins replied that the problem seemed to be, in part, concern by the company over the reaction of the unions and the public. Although 5 percent may be a small number, it can still be perceived as a example of moving jobs overseas, even if that process may have been responsible for getting new work. In addition, the prime contractors were concerned about revealing contractual information in the report.

Technology Transfer

A second policy issue raised by Dr. Watkins concerned technology transfer. He stressed that the most important technology transferred in this case of a mid-tier, metal-banging company was tacit manufacturing knowledge. GAMC was not transferring design know-how or proprietary in-house sophisticated manufacturing technologies, as this was a mature, mid-technology product. They transferred manufacturing know-how that allowed the foreign supplier to move more rapidly down the learning curve. It was experiential knowledge, such as engineers noticing that a pneumatic press is turned up too high. They were also training their foreign suppliers in their customers' preferred procedures, such as their total quality management process.

This created a sensitive issue for the company. With Japanese engineers visiting the U.S. production facilities to learn how to make the product, managers needed to acquire the new skill of dealing with workers worried about their jobs being transferred offshore.

Dr. Watkins pointed out that this type of technology transfer is a nonmarket behavior. It is the type of activity that goes on routinely within any company. It is difficult to police this from a policy perspective. Nor can it be distinguished from the classical reasons for strategic alliances, which is to learn from one another and thereby transfer tacit knowledge.

Trade Distortion

The case study illustrates that trade distortion, while occurring, may be inevitable. The agreements were not necessarily based on price. GAMC transitioned products to the United Kingdom, South Korea, and Australia because these are big markets for their customers, not because of the price differential. Dr. Watkins also noted that the units being supplied to GAMC from foreign sources were going into all their customers' aircraft, not just those for Japan or Korea. There is not a specific offset for a specific customer. The outsourced production goes to all customers. The result is a marginal increase in trade distortion.

There is also an economic distortion in the fact that the process has increased the burden on GAMC. Their costs and responsibilities have increased because they now have to manage a process of seeking out and training international sources. Overhead costs, especially the management of procurement, have been rising as the top tier has been pushing this responsibility down the supply chain.

Market Access

These arrangements are made for market access reasons, where governments have put pressure on foreign aircraft manufacturers. Offsets are not the direct problem. The problem is the trade-restricting policies, which offsets seek to get around. Companies, especially middle-tier companies, can do little about offsets. The target of the U.S. government should be the trade-distorting policies.

In trying to apply trade law to offsets, the measurement problems will be enormous. The case has shown that much of the problem will remain hidden because it is so many layers deep in the supply chain.

The case also illustrates that the process is unlikely to create new competitors for the top tier with the transfer of tacit manufacturing knowledge. The barriers to entering the top tier are too great. However, this process is expanding the foreign skill base. Nagoya got work directly from Norton, bypassing GAMC in part because of the general "introduction" made by GAMC. GAMC claims they would not have gotten work anyway, but Dr. Watkins believes they had the skills in-house to do the job. The result of the technology transfer is not to turn these foreign companies into competitors, but to help them get down the learning curve faster then they could on their own.

In closing, Dr. Watkins pointed out that companies are facing what he calls "mid-tier squeeze." Top-tier firms are passing responsibilities down to the lower tiers, teaming with the lower tiers, increasing the risk and overhead of the lower tiers, expanding the necessary set of skills in the lower tiers, passing down responsibility for design and risk sharing, and demanding foreign sources. The mid-tier firms follow the same lean manufacturing paradigm of attempting to work with the best suppliers and make them better. The greatest danger to existing mid-tier firms is that of fostering "tier jump," for example, helping a lower-tier firm become a competing mid-tier company. Mid-tier companies are under enormous pressure to take more responsibility and are, at the same time, creating their own competitors.

Dr. Wessner asked how there can be "tier-jump." Would not the American prime contractor ask Nagoya to train GAMC in return, perhaps with the encouragement of the U.S. government? Dr. Watkins responded that the reality is that mid-tier firms do not have a lot of bargaining power, nor is there any incentive for the American prime contractor to make such a demand.

In response to a question from Joel Yudken of the AFL-CIO as to how typical the situation facing GAMC is, Dr. Watkins stated that his impression is that mid-tier firms are under increased pressure from the upper tier.

DISCUSSANTS

John Sandford
Rolls-Royce, N.A.

Mr. Sandford opened his remarks by commenting that he believed that the case was very representative of what is happening in the industry. The reason the prime contractors push their suppliers is because the prime contractors have market power and access to capital. From the perspective of a CEO, he stressed the importance of moving beyond technology to looking at the extremely crucial role of capital. He has seen numerous cases of prime contractors putting costly pressures on mid-tier firms. But he has also seen cases in which the prime contractor has subsidized lower-tier firms in order to meet their obligations. Although this may be recorded as an offset, the company is getting the capital it needs. He reminded participants that companies need to satisfy the demands of their shareholders. Mr. Sandford was especially pleased that this case study described the real costs. Another example of costs is the process of gaining Federal Aviation Administration (FAA) certification. In the production process described in the case, parts probably have to travel across the Pacific at least twice. All of the steps have to be approved by the FAA because the certification states that the parts have to be "as designed," "as manufactured," and "as proven." This is an additional cost.

Mr. Sandford also stated that, from personal experience as a chief operating officer, the overhead costs of managing the process are very real. However, the companies have no option. In the long term, companies may be training their competition. But, given that the process may take ten years, companies have time to decide where to move. Although the cost of entry is enormous, there is talk of aspirations of companies in China and Japan to become prime contractors. Mr. Sandford stated that in his opinion no sane person would undertake these costs, but some companies and their governments may do it nonetheless. In any event, creating a mid-tier industry creates the ability and the political leverage for a country to move into the prime tier.

In summary, Mr. Sandford believes that the analysis presented in the paper was extremely realistic. Although it is unclear whether it was generally right or wrong, GAMC probably made the right decision from the CEOs' and shareholders' perspective. The company has ten years to make their next strategic move, and they did get the current business in the meantime to satisfy short-term shareholder demands.

Al Volkman
U.S. Department of Defense

Mr. Volkman stated that he would focus his remarks on the issues covered in the paper of dual-use management and strategic international sourcing rather than offsets specifically. These two issues are similar to those identified by the Undersecretary of Defense for Technology and Acquisition as key issues facing the U.S. industrial base: commercial–military integration and international armaments cooperation.

The Revolution in Military Affairs

The security environment facing the United States is one of a changing and unpredictable threat. One of the key factors is the development of new technologies. The new phrase is "the revolution in military affairs." This refers to the increasing importance of communications and information technology and biotechnology in the national security structure and in the way in which wars will be fought in the future. It is important that the United States maintain access to these technologies to maintain a strong defense. In this environment, it is important to bring equipment into place faster and cheaper.

The United States, in the future, will be forced to place a greater reliance on coalition warfare. Almost every major war fought by the United States in this century has been fought with allies. These coalition operations will be even more important in the future.

Other factors affecting the security environment include

- the fact that DoD has a smaller share of the national budget;
- the increasing globalization of business activity; and
- the fact that the commercial sector, rather than the defense sector, is increasingly driving the pace of technological change.

Less Time and Lower Costs

The DoD needs to make sure that suppliers in the twenty-first century are able to design and produce the highly complex weapons systems necessary for national defense. The DoD believes that it takes too long to field weapons systems, and that the cost of the systems needs to come down. In addition, these systems must be designed to support coalition warfare and interoperability.

Although commercial technologies are driving the revolution in military affairs, there is also a need for revolution in business affairs. The DoD needs to look at ways to improve its acquisition processes. Important advances in this area have been made. As the paper points out, the defense business operates on military standards and specifications for contracting with companies. In the past, there have been some suppliers, such as Hewlett-Packard, who did not want to do business with the military because of these requirements. Over the past few years, the DoD has tried to rely more on commercial standards. Commercial products are a bigger and bigger part of the procurement system.

Given the globalization of the commercial industrial base, there will be more foreign components in our systems as DoD becomes more dependent on the commercial sector. It is therefore important not to impose restrictions, for protectionist reasons, that would limit the military's access to those items.

International armaments cooperation between DoD and U.S. allies is fundamental to coalition warfare in the future. Cooperative development and production of major weapons systems have not been all that successful in the past. Such cooperation is not popular and has generally been resisted by Congress, unions, and much of the military. There is an unhealthy trend toward both Fortress Europe and Fortress America.

Coalitions Mean Interoperability

Nevertheless, globalization of the defense industry and coalition warfare is the direction of the future. Coalition warfare requires interoperability; interoperability is greatly aided by operating the same equipment. Likewise, it is important for U.S. allies to be using the highest quality equipment on the battlefield in support of U.S. battlefield operations. At a time when the defense budget does not match the military's modernization needs, it is important to leverage the investments that Allied governments have to put into weapons systems. European governments are no longer able to develop highly expensive weapons systems on their own. The costs of U.S. programs, such as the Joint Strike Fighter, have reached the point where financial support from U.S. allies might be welcomed.

Mr. Volkman closed by arguing that, to ensure the success of cooperative programs, we must first look at what U.S. warfare needs are. Next, it is important early in the cooperative programs to agree on the business rules and agreements to share technology. Governments will need to make broad agreements on the relative contributions by each government and how much work will go to each country. Companies can then form teams and decide how these rules will be applied. Such an agreement guaranteeing that the United States receives benefits equal to its contribution is crucial in building public support for such cooperative programs. Already, Congress requires that funds for NATO R&D cooperative programs be spent in the United States. That may be trade distorting. It is clearly a fact of life.

GENERAL DISCUSSION

Art Ismay, Defense Industry Offset Association and Rockwell International: Mr. Ismay commented that the case very well described real life in both the aerospace industry and the automotive industry. However, the process is not just dictated by customer demands. The manufacturers themselves undertake strategic outsourcing to improve quality, improve delivery, and reduce costs. Cost saving is achieved by selecting the most competitive supplier available. The lowest-ranked suppliers, who are not good with delivery, are poor in quality, and are not measured well in cost, will be dropped. Companies will work with suppliers to simplify the production process to achieve cost savings. Take for example a product that requires two operations, such as forging and machining. Companies will help upgrade suppliers so that they can do both operations, thereby reducing the number of parts of the process that have to go out for separate bids and thus reducing the overhead. Companies will also send personnel to suppliers to upgrade their processes because the companies need results immediately.

Sometimes outsourcing is done for purposes of market access. But companies would not outsource to their disadvantage—they would rather walk away from business. Mr. Ismay emphasized that supply chain management is done to benefit a company, not to benefit a foreign government or for the sake of completing an offsets program.

Mr. Ismay disagreed with Dr. Watkins' concern over "tier jump." The work is parceled out in such a way as to minimize this. Companies work to upgrade their suppliers' abilities to satisfy their needs (e.g., better gears). They do not help them produce higher-tier products (e.g., a 16-gear transmission).

Deborah Nightingale, Massachusetts Institute of Technology: Dr. Nightingale emphasized her agreement with Mr. Sandford as to the importance of capital.

Decisions are made based on capital requirements. If you do not have the money to develop a new, expensive product, your options are to not do it, or to find partners. Access to capital is essential to grow the business and increase productivity; increasing productivity is easier when business is growing rather than downsizing. Growing the business in aerospace means expanding internationally. That requires partnerships, which are a way of life in many industries other than aerospace, such as automotive and electronics. They require highly skilled people to manage the technology so that you do not give it away. International partnerships may be difficult, but they are a way of life in the globalized economy.

Emerging Challenges and Diverging Interests

Kirk Bozdogan
Massachusetts Institute of Technology

Dr. Bozdogan presented findings from a recent study by the Massachusetts Institute of Technology (MIT) Lean Aerospace Initiative. He emphasized the complexity of the issues and the difficulty of measurement and interpretation. Although some believe that there is no need for a specific policy on offsets, the offsets issue relates to a number of larger issues that do need to be addressed. In attempting to take up these issues, he noted that his presentation would take both a historical perspective and a look at the future. He proposed to raise questions and move toward answers, noting that, at some point, we must stop talking about the issues and provide solutions.

Current Trends

The aerospace industry has undergone a process of consolidation, restructuring, and downsizing over the past decade. This process was driven by a downturn in the commercial aircraft industry and by deep cuts in the defense sector. Although important, the impact of offsets has been dwarfed by these larger factors.

At the same time, there has been a clear shift in the innovation model in the aerospace industry. Government's role has progressively diminished. R&D support for the industry has declined, as has infrastructure such as wind tunnels. However, the change in the structure of the innovation process is a consequence, and not a cause, of the larger changes in the industry, as will be discussed later.

Dr. Bozdogan agreed with previous speakers as to the trends in the industry. Growth in overseas demand has outpaced domestic demand. Although U.S. markets have dominated, the main growth is now occurring overseas, especially in the Pacific Rim. U.S. dominance in commercial aircraft production has been challenged, but not overturned. International collaboration and subcontracting has increased the globalization of the industry.

On the defense side, concerns over affordability have dominated. The industry has focused on the reduction of cycle times and costs. In fact, these were the reasons for establishing the Lean Aerospace Initiative at MIT. The Initiative is a consortium bringing together industry, government, and unions to achieve significant fundamental improvements in the way in which weapons systems are designed and built. The Initiative collects and analyzes best practices from around the world, together with a research agenda directed by consortium members. The results are disseminated straight down the supplier chain.

The defense sector is also the focus of a drive toward greater utilization of commercial technologies and practices, as previous speakers mentioned. This includes utilizing the commercial industrial base to meet military requirements.

Fewer Suppliers

Dr. Bozdogan then presented the results of a survey of the aerospace supplier base, broken down into three sectors: airframes, electronics and avionics, and engines and other. The survey found a dramatic decrease of over 50 percent in the number of direct production suppliers between 1991 and 1995 in all three sectors of the industry. The story is the same in both the commercial and the defense sides of the industry.

Higher Certification

Although companies are decreasing the number of suppliers, they are increasing the certification requirements of the suppliers who remain. Certification is being used to ensure that suppliers are fully qualified in terms of process technology. This is part of the larger issue of how companies synchronize their production technology with that of their suppliers. The result is that companies are able to devolve more responsibility to their suppliers, including design, development, inspection, and testing. In response, suppliers are taking a larger responsibility for more complex subassemblies. This requires a carefully forged long-term relationship between companies and their suppliers.

More Information Sharing

A key part of these long-term relationships is the sharing of information. Within the aerospace industry, there has been a significant increase in communication and information sharing between companies and their suppliers. Included is information as to production costs, the supplier's statistical process control processes, and their performance improvements. This represents a sea-change in the relationship between companies and their suppliers.

More Outsourcing

The survey shows that prime contractors are undertaking outsourcing for a variety of reasons. Chief among these is lower costs. Ninety percent of the companies responded by saying that lower cost was the reason for the outsourcing of the largest dollar-value products that were previously built in-house. The second major reason given for outsourcing was the strategic realignment of production. This reflects the structural shifts within the industry, the slowdown in commercial markets, the cutbacks in defense spending, and the streamlining of production operations. In essence, companies have reassessed their core competencies as to what they should build in-house and what they should devolve to suppliers in order to make large efficiency gains. Higher quality was also given as a reason for outsourcing, although this was less significant than other factors.

Production Partnerships

Two types of long-term partnerships have evolved in the industry. One of these Dr. Bozdogan referred to as "production-focused supplier partnerships." These involve support given to suppliers for ongoing production operations. Eighty-five percent of firms in the industry have established this type of long-term purchase agreement with their key suppliers. The major reason given was to reduce costs, followed closely by a desire to minimize future price fluctuations and to realize mutual performance improvements. It is important to note that these are truly mutual improvements, not simply companies demanding improvements from their suppliers. These long-term agreements have taken a number of forms, including partnerships focused on one or more products for more than three years, multiyear design/build teams, and ongoing partnerships.

Development Partnerships

The second form of long-term partnership Dr. Bozdogan referred to was "development-focused supplier partnerships." These partnerships focus on product development, technology development, and process development. Over half of the firms in the industry have established these development-focused supplier partnerships. One of the main reasons given by companies is to win new business—in essence, market access. Another reason was to enhance supplier performance capabilities—companies helping to improve the suppliers. Also important was achieving material cost reductions. Eighty-six percent of the respondents believe that these partnerships are either "very important" or "absolutely critical" to their long-term competitiveness.

At the same time, there is evidence of a shift in the industry from partnerships to knowledge integration. In a way, the aerospace industry is replicating the experience of the electronics industry, where companies are engaging their suppliers in the front end of the R&D process, rather than only at the back end of procurement.

The old model featured top-down control with the interfaces between prime contractors and suppliers totally defined and well controlled. Product requirements were passed straight down the supplier chain. The current lean model features more collaboration and partnerships, but still retains existing organizational boundaries. There is collaborative information exchange and some risk sharing, but these activities are constrained by pre-established contractual arrangements.

Virtual Teams

The emerging model is one of "virtual teams." These teams of prime contractors, key suppliers, and sub-tier suppliers operating without boundaries have fostered architectural innovation that has resulted in significant benefits in cost reductions, cycle-time reduction, and quality improvements. Under this new model, best practice is when the interfaces are allowed to freely change to optimize the system. The enablers of this new system include long-term commitment, open communications, and maintaining trade secrets to allow such communications to occur.

The emergence of these closely linked partnerships and networks has changed the industry, to the industry's benefit. The industry now has one common and highly integrated domestic supplier base supporting commercial and defense activities. It is crucial to maintain and strengthen that supplier base to support production and maintain the existing aircraft fleet.

Risks Ahead

Looking ahead, opportunities and demand for offsets will accelerate. One potential risk to the U.S. supplier base is excess capacity in defense production. To the extent that offsets create new foreign competitors, cost and competitive pressures on the U.S. supplier base are exacerbated. Displacement of U.S. suppliers in specific areas is a real possibility. Dr. Bozdogan reminded participants that other governments are actively attempting to boost their own nations' aerospace capabilities.

Another risk to the industry concerns the nature of the new production model. This new model of collaborative partnerships is more conducive to technology transfer. As the United States moves toward this new model, there is the potential for an exodus of U.S. technology overseas.

Dr. Bozdogan referred to the questions raised in his paper. In particular, he remains concerned as to what are the long-run national security implications of greater military reliance on the commercial industrial and technological base as that industrial base becomes increasingly internationalized. The situation can be characterized as a prisoners' dilemma, requiring negotiations with our allies. But we cannot count on the "invisible hand." We may not know where our suppliers are when we need to engage the system's surge capability. The risks facing the industry are very real.

DISCUSSANTS

William Reinsch
U.S. Department of Commerce

Undersecretary Reinsch began by commenting that he found Dr. Bozdogan's analysis very useful. He noted that he would confine his comments to the questions and issues raised in the presentation. Specifically, he commented on a similar situation concerning the national security implications of the military moving to the commercial technological base as pointed out in the presentation. The technology involved was high-performance computing.

In the high-performance computing sector, the normal government procurement process takes longer than the normal product life cycle of the computer. If the government uses normal procedures, it will always be technologically behind. Thus, the Pentagon is moving to commercial technology and commercial procurement in computers. The military has discovered that it relies on high-performance computing for a variety of applications, ranging from those involved in direct conflict activities to applications such as real-time battlefield weather prediction.

They have also discovered that they do not buy enough to keep any of the small companies that make high-performance computers in business. The military sector simply is not a significant enough buyer of a civilian technology to make much difference to the health of a sector. The DoD needs the companies to be healthy in order to "run faster" to maintain a lead in the technology. Exports contribute significantly to the health of the industry, but exports mean the globalization of the industry. This raises a number of national security problems because globalization means the handing over to others of an important military capability, either in the form of the export of a product or in the form of technology transfer. Exports, therefore, end up defeating the purpose of the U.S. military in maintaining dominance in this key technology. The dilemma facing the DoD is that if you do not allow exports, you endanger the viability of some of the companies who produce directly for defense use.

After grappling with the issue, the DoD decided that it was in their greater interest to maintain the health of the industry and deal with the technology risks in some other way, rather than try to keep the technology within the United States. This illustrates the problem facing other sectors as they go global.

In response to the question raised in the paper as to whether trade policy and technology policy know what each other is doing, Undersecretary Reinsch stated that in his experience, the answer is no. This is a chronic problem both between agencies and within agencies, as it is within Congress. There are trade people and there are technology people. They each have their own portfolios and mind-sets and tend not to talk to one another. Thus, policy tends to end up disjointed.

Undersecretary Reinsch went on to point out that, clearly, corporate business interests and national strategic interests can and often do diverge—collectively and individually. Companies enter into business arrangements that may be in their interest, but raise significant questions as to whether the activity is in the national interest. This raises questions about the process.

A related question concerns how to look at the health of an entire sector, such as aerospace. Are the activities in a sector convergent with both business and national interests? Specifically, are the developments in the aerospace industry, which might be very much in the interests of members of the industry in terms of profits and jobs, in the national interest and the national security interest? The answer is not necessarily yes. This poses a difficult question. The companies are doing exactly what they are supposed to—they are making money to presumably invest in the next generation of technology. That activity, however, may run up against national security goals.

There is no process in the government to address these questions, except in isolated circumstances. There is a process concerning the review of foreign acquisitions of U.S. companies. But that is not a process where these issues are frequently addressed. There is no process to regularly address these questions that are raised in the process of U.S. companies globalizing, other than through ad hoc arrangements.

There is an attempt to address some of these issues in the intergovernmental Committee on National Security as part of the National Science and Technology Committee, which is chaired by the President's science adviser. The original purpose of this committee was to look at gaps and redundancies in government R&D. The committee has taken on as an additional task the process of trying to identify a way in which government can more systematically address these globalization issues on an interagency basis. There is no desire, nor is there legal authority, to intrude into private business transactions. But there should be someone who speaks for the national security interest—even if that role is only to publicly point out that interest.

The appropriate government role may be simply to point out how these events impact on national interests. There is no governmental authority to go further. There is authority to control technology transfers for national security purposes, but not for competitiveness consequences. National security is not defined in a broad manner so as to encompass the issues being discussed at this symposium. This interagency committee is looking at ways to address these issues, and if it succeeds in developing a process it will surely consult with industries such as those represented at this meeting.

Albert Kelley
Massachusetts Institute of Technology

Dr. Kelley noted that offsets evolved from a decision years ago at the White House to promote foreign arms trade as a reaction to what other countries were doing. Offsets were devised as a mechanism to provide financial inducements to other countries that were legal under U.S. "corrupt practices" standards. This mechanism was not devised as a complete set of activities known as offsets. Rather, it grew out of a series of unique individual deals.

Dr. Kelley noted that he was always worried that these mechanisms left U.S. companies naked to face off against foreign governments. The policy of the U.S. government has always been "we do not like offsets as a government, but, on a private basis, go ahead on your own." Companies have done a good job in handling the issue in a mature and ethical way.

Dr. Kelley went on to raise a number of "yellow flags"—issues to worry about. Buyers want an economic return from offsets, more than simply the product. This means that various ministries, such as the finance ministry, get involved. The result is a greatly complicated process. It is no longer a straight defense-to-defense airplane deal. The process becomes one of barter—trying to determine what type of offsets will be involved. He used the analogy of leasing an automobile, where there is a whole series of mini-offset agreements concerning interest rates and dealer discounts.

Defense buyers often want to add capability to the products they are purchasing. They want to have better technology and capabilities more than anyone else—even if it is just for show. Sometimes they are encouraged to do this by the contractor. This leads to a general escalation of military requirements, because no one, including the United States, wants to be number two. A case in point was the F-20. It is important to stick to the purpose of the weapons system—the core military requirements. It is also important to avoid having countries buy the wrong plane because of the offset. There have been deals in which the country was more interested in the offset package than the airplane.

Dr. Kelley stated that it is a myth that foreign sales reduce the cost of airplanes by increasing production. As was pointed out earlier, requirement for the recoupment of R&D costs is often waived. In addition, orders from foreign governments are different enough so as to essentially require the building of custom planes. Unlike the automotive industry, every plane in the production line is a different airplane. Thus, foreign sales do not result in significant savings due to greater economies of scale.

The key concern, including that of the U.S. government, is jobs. Dr. Kelley noted that the Apollo space program was undertaken, in part, to create jobs in the aerospace industrial base that was declining after the Korean War.

He closed by noting that coalition warfare and the U.S. ability to ramp up defense production very quickly are important keys to national security. Reform of the acquisition and procurement system will be required to accomplish this, including more sole-sourced contracts. Developing ties with foreign countries, including the use of offsets, can help strengthen the U.S. defense industrial base. Offsets are not bad per se, but there are bad offsets. A solution to the issues raised by offsets will undoubtedly develop, so we should not kill off offsets before we understand their consequences.

Panel 7

The Role of the U.S. Government in Setting Offset Policy

Owen E. Herrnstadt
International Association of Machinists and Aerospace Workers

Mr. Herrnstadt noted that participants, as was the case during the June 1997 workshop, were grappling with the definition of offsets and the difference between direct offsets, indirect offsets, and voluntary marketing schemes. Rather than use these narrow definitions, he suggested that participants think of offsets in broader terms to cover all types of performance requirements as suggested in Dr. Mowery's paper. For policy reasons, it may be important to make such distinctions. For real workers, the distinction between each of these different types of offsets is nonexistent. A layoff is a layoff. He urged participants to always keep in mind what the practical effects are on workers and sub-tier producers.

With this in mind, he outlined areas covered in the paper where there is some common understanding of the issues. First, offsets are increasing—although there might be disagreement as to the rate of increase. The Commerce Department report shows indirect offsets increasing more quickly than direct offsets. Voluntary marketing strategies and agreements, which are really not well understood, also appear to be increasing. This observation is based on the increasing number of such agreements reported in the trade journals. Offsets and similar types of marketing strategies have become the norm and are viewed as a necessary evil.

We also know that we do *not* know the broad extent of the impact of these agreements on the U.S. labor market because of a lack of reporting requirements. Although there are some reporting requirements on the military side, reporting requirements in the commercial aerospace industry are woefully missing. In addition, the effects on industries outside of aerospace have yet to be reported.

We further realize that the outlook for aerospace employment is not promising. Earlier in the symposium, Dr. Scott reported that up to 25 percent of the production jobs in the industry could be lost by the year 2013. There are other studies that indicate that offsets continue to have a negative, and possibly increasingly negative, impact on aerospace production workers. Although some note that currently there is job growth, the current economic boom could turn down rapidly as a result of the Asian crisis. In addition, the gains made in the past few years still have not compensated for all the jobs lost earlier. The issue of employment—and the pain of dislocation—is still very real for those workers who have lost their jobs or who have found new jobs in other industries at lower wages and with fewer benefits.

As Dr. Rosen pointed out earlier, the aerospace industry is important because it makes a large contribution to the U.S. economy. The administration understands the importance of preserving and assisting the industry. Maintaining the health of the aerospace industry is clearly in the national interest.

Certainly other countries believe that having an aerospace industry is in their national interest. Countries such as Japan, China, and the Western European nations, have well-defined offset policies. As NATO expands, Eastern European countries are now engaging in the offsets game. Many of these countries are using offsets to develop their aerospace industry, in particular China.

Multiple Effects

As a result, there is concern over the widespread employment effects of offsets on both the prime contractors and the subcontractors, as well as on those who are caught in the draft of indirect offsets. Technology transfer is also a concern, including the possibility of taxpayer-funded research being shipped abroad to increase the strength of our competitors. As mentioned earlier, this simply worsens the problem of global overcapacity. National security concerns are also important. It is very hard to distinguish when a technology is transferred to a country such as China whether that technology will be contained in the commercial sector or used for military purposes. This was especially well described in a presentation at the June 1997 meeting.

There is a concern regarding prime contractors versus sub-tier suppliers. A presentation earlier in the symposium described what is happening to the aerospace sub-tier industries. The process of sub-tier companies being squeezed out of the business has a real effect on jobs. Mr. Herrnstadt then read three comments from sub-tier producers quoted in the most recent Commerce Department report on offsets as to the negative impact of offsets on their business and employment.

No Governmental Coordination

In addition to conflicts between private sector entities, there is a lack of coordination between public entities. For example, the FAA developed a proposed regulation concerning fees on the production of "complex parts and subassemblies outside of the United States." One of the purposes stated for the proposed regulation was to take advantage of lower labor costs, lower manufacturing costs, and "fulfilling certain aircraft purchasing requirements that require a production approval holder to produce a percentage of the aircraft within the purchasing country." This sounds like offsets. There is a real question as to whether or not this type of move would have a significant effect on certain other entities or members of U.S. society. Mr. Herrnstadt said he believes it would have such an impact.

Three Types of Solutions

There seem to be three categories of solutions represented at this symposium. The first is the "do nothing is best" model. This is analogous to a pilot seeing thunderstorms ahead on the radar and turning off the radar to solve the problem. The second category seems to be a recommendation for more ad hoc meetings and a call for more information. This is the approach taken recently by the Trade Promotion Coordinating Committee. Although welcoming the acknowledgment of the lack of data, Mr. Herrnstadt argued that more needs to be done.

He urged a third solution—-a framework to begin addressing these issues. A starting point for such a framework, which should be quickly put in place, would be a mechanism for acquiring information on offsets through regular reports. Although there is already some information required on the military side, the information is insufficient to help us understand what is happening on the sub-tier level. There is virtually no information on these issues as they affect the commercial aerospace industry.

In closing, Mr. Herrnstadt noted that his paper contains additional, specific recommendations. In particular, he stressed the need for additional funds for trade adjustment assistance. Reminding the participants that the stakes are growing and that the pieces to the offset puzzle are becoming more difficult to identify, he argued that the United States can no longer sit back and let other nations, and the hundreds of private parties involved in offsets, determine our future. It is time for the U.S. government to take a strong leadership role in developing a policy on offsets and related employment and trade issues.

DISCUSSANTS

Greg Martin
Lockheed Martin

Mr. Martin began by agreeing with Mr. Herrnstadt and other speakers that competition has driven offset demands to increasing levels, both with respect to size and scope. In the defense industry, this has been the result of the dramatic reduction in U.S. government procurement and the accompanying increase in competition for foreign sales.

Foreign governments are spending more and more taxpayer money on defense acquisitions from outside their borders. They need to justify these expenditures not only to their taxpayers but their own labor unions, because they are shipping jobs offshore. In many countries, labor unions are driving the offset requirements. If the shoe was on the other foot and the United States was purchasing the bulk of its defense products from overseas, U.S. taxpayers, defense companies, and labor unions would all be crying for strong offsets. The United States does have an offset program in the form of the "Buy America Act," which requires that at least 50 percent—and typically a much larger percent—of any defense products purchased offshore be manufactured in America. Neither defense companies nor unions oppose the "Buy America Act" because it does create jobs and business here in United States.

Mr. Martin pointed out that there was a tremendous employment gain in the 1980s due to the defense buildup. Unfortunately, that level of employment was unsustainable. With the end of the Cold War and the downturn in the commercial sector in late 1980s, employment fell. Offsets had a relatively minor impact on jobs. Some of the largest offset programs occurred during the 1980s while employment was still growing. Thus, critics cannot have it both ways by claiming the downturn was due to offsets but the upturn in employment was not.

The defense sector today depends primarily on international sales to keep production lines open in a number of products, such as the C-130J, the F-16, the F-15, and a number of helicopters.

Although understanding the concern over a lack of data, Mr. Martin urged participants not to minimize the significance of the amount of data that is already being collected. Defense companies are providing a tremendous amount of detailed proprietary, sensitive data. It is a substantial task for the companies to put this data together and supply it to the Commerce Department.

He agreed that there will be aerospace jobs lost in the future as the industry continues to consolidate. The consolidation is the result of market reality, not a result of offsets. He disagreed with earlier analyses stating that two-thirds of the jobs would be lost because of market access problems and one-third lost because of offsets. The key question is, if you do away with offsets, what happens to the market access?

He expressed concern about the Commerce Department's use of anecdotal information regarding the effects of offsets on subcontractors. The government goes to great lengths to get hard data from the defense prime contractors. It is thus disconcerting to see the report's findings based solely on unsubstantiated comments from subcontractors. The industry recommends that the Commerce Department change its methodology of collecting these data to find out who is really being hurt by offsets and to what degree. The comment from suppliers came at a time when the industry was consolidating and rationalizing the supplier base. Some of these suppliers may have been dropped because of that consolidation, not because of offsets.

Industry supports multilateral initiatives, while not holding out a lot of hope that these initiatives will be successful. However, any initiatives and policies must not impede in any way industry's ability to form international relationships.

Mr. Martin closed by noting that data are being collected and a dialog is continuing. Industry supports both the collection of data and the dialog and is happy to participate in developing a plan to go forward with multilateral discussions.

Thea Lee
AFL-CIO

Dr. Lee commented that she thought Mr. Herrnstadt's paper was not only a persuasive case for the need for action but also a forceful call for a strategy and a coordinated set of initiatives. The first point Mr. Herrnstadt made is that the problem is severe and growing. Problems include the effects on workers and suppliers as well as the future impact of offsets packages. The second point is that there is no other player who can address these issues other than the government.

Different Perceptions

Industry clearly believes that offsets are not a major problem. It is not surprising that the issue is perceived differently by industry and workers. It is an issue of inconvenience to industry. It is a problem of much greater significance to workers. Industry can move production to gain market access and still earn a profit; workers are left out. Dr. Lee remains unconvinced by Mr. Martin's argument that the existence of offsets during a time of employment growth in the 1980s means that offsets had no negative impact on employment. One must look at the impacts of offsets, other things being equal. It is hard to argue that, when companies are forced into offset deals, there is not a negative impact on at least potential employment. This is a different point than looking at aggregate employment. She commented that Mr. Herrnstadt's paper could emphasize this point by analyzing what would have happened to employment in the absence of offsets.

Counterbalancing other Governments

Dr. Lee then went on to the issue concerning government actions. The argument is made that U.S. companies must accept these deals imposed by foreign governments in order to get the business. If they are not allowed to make these arrangements, they will lose out to companies from other countries. This is correct only in the absence of some type of U.S. government action. You cannot counter a nationalistic industrial policy with a laissez faire approach. This resulted in a mismatch. You will either have to let market forces work and gain the efficiency benefits or take government action in the presence of these interventions by other governments. This can be done either by negotiating multilateral rules or by countervailing.

Multilateral Rules?

Dr. Lee added that it was heartening to hear industry representatives state that they would support multilateral rules. However, without more energy behind the push for multilateral rules, there will be little incentive to undertake such negotiations. And in the absence of such multilateral rules, the United States should be willing to adopt the second-best policy of taking defensive unilateral action.

A Commission Needed

The policy recommendations described in Mr. Herrnstadt's paper are both modest and clear. A good first step would be the creation of a government commission that can gather information, coordinate different areas such as trade and technology policy, and develop guidelines for multilateral rules.

Dr. Lee closed by commenting that we still do not have good information. We need a catalog of offset deals and their impact on the supplier industry. Everyone agrees that we do not know how much subcontractors are hurt by offsets. Although some information is out there, the right information is not yet available. Gathering that information is crucial for making informed decisions. She commended the National Academy of Sciences for bringing together the interested parties at this symposium—and at the June 1997 meeting—as part of the first step in reaching consensus on this very real problem.

GENERAL DISCUSSION

Brad Botwin, U.S. Department of Commerce: Mr. Botwin noted that the Commerce Department's Bureau of Export Administration, where he is the director for Strategic Analysis, appreciates the efforts of the prime defense contractors in supplying the information. However, the information given to the U.S. government is only a fraction of what the companies give to foreign governments on how they have fulfilled their offset obligations. The Commerce Department would be happy to simply receive the same information given foreign governments. Concerning subcontractor data, the Commerce Department is updating and upgrading that information, including a forum in Austin, Texas, on April 1, 1998, to discuss the impact of offsets on subcontractors. Each of the companies quoted in the report are legitimate, high-technology, financially stable suppliers to the U.S. industrial base. The companies had asked that their names not be used out of fear of repercussions. He mentioned that the pain being inflicted on the supplier base is very real.

Sally Bath, U.S. Department of Commerce: Ms. Bath noted that the industry will face a great deal of change in the future, including changes that we cannot anticipate and are not prepared to address adequately. Offsets may be last year's problem. European targeting of the supplier base, particularly avionics, may be only the tip of the iceberg. We need to look carefully at the manner in which we maintain an industry that everyone agrees is essential.

Art Ismay, Defense Industry Offset Association and Rockwell International: Mr. Ismay commented on the general concern expressed about the lack of real data as to the impact of offsets on employment. Obtaining this information would require looking at proprietary data. The government does not need to ask for sales contracts. The data provided foreign governments give the specific details of the offset contract and how well the company is doing in meeting those requirements. This is proprietary data that companies are reluctant to have revealed to their competitors. Companies would supply this information, if required, but would strongly resist having such requirements. In addition, the prime contractors do not necessarily have, nor can they get, information as to how the offset requirements affect employment among the subcontractors. Labor unions might be better off coming up with some scheme to obtain the information more directly than imposing a requirement on the prime contractors.

John Shaw, Cambridge Consulting Group: Mr. Shaw noted that there is no consensus as to whether the entire phenomena of offsets is good, significant, or insignificant. Government policy is made in a political context. Data are only as good as the interpretation. No matter how much is collected, the information will still be used to make the most convenient interpretation. As discussed earlier, perceptions may be more important and bad data may push out good data. There will never be enough information in the U.S. government to have a finality to satisfy all the participants. It is a resolution problem, not a data problem.

Randy Barber, Center for Economic Organizing: Mr. Barber agreed that there is a problem with interpretation. However, the lack of data reinforces suspicion about the problem. The fundamental argument from industry is that, on net, offering these types of offset arrangements creates more jobs than if they were not offered. The problem is that there is no way to verify this. Proprietary information is a legitimate concern. However, the industry needs to balance whether it is more damaging to have unfounded suspicions and an information vacuum, or if it is better to have some basis for making a judgment. At present, we lack the information needed to determine whether these deals, which basically trade production for market access, adequately take into account the interests of workers, national security, and the broader economy. Information is the precondition. The fundamental question is whether these deals result in a net creation of jobs. The lack of information means we are trying to answer that question in a vacuum.

Greg Martin, Lockheed Martin: In response, Mr. Martin cited the example of the F-16 line, where presently less than 10 percent of the production is for the U.S. government. He asked what type of data would convince Mr. Barber that offsets are not saving those jobs. Mr. Barber replied that may very well be the case in that particular example.

Steve Beckman, United Auto Workers: To put the offsets issue in perspective, Mr. Beckman suggested that we consider the difference between the auto and aerospace industries. U.S. auto companies are international in scope, but they do not serve those foreign markets through exports. U.S. auto companies moved production facilities overseas because of offset requirements placed on them in the 1920s and the 1930s. The ability of the auto industry to provide decent paying jobs for workers was never dependent on those exports. The companies prospered as their international sales prospered, but employment in the industry was based on its domestic strength. This form of national economic development was the norm in the first half of the twentieth century.

Now, the norm is export-led development. The aerospace industry has grown up in a time when its success depended on concentrated domestic production and the export of that production to serve foreign markets. A process is now being imposed on the aerospace industry over a period of 10–15 years that had been imposed on the automotive industry over a period of 20–30 years. Workers who lose their jobs in this process of internationalization are expected to think it is a great idea because it is better than if industry moved totally overseas to produce for those overseas markets.

The reality is that production is moving offshore because there are barriers to market access, not because U.S. products are noncompetitive. Yet workers are being told that they have to be competitive to succeed internationally. That is why workers do not trust what the companies tell them about offsets. Workers are going to lose their jobs and companies are going to keep the profits. Free markets, we are told, are the solution to our economic problems. But if these markets are not open, we cannot continue to play the game as if they are—and that is what workers are being asked to do.

Owen Herrnstadt, International Association of Machinists and Aerospace Workers: Mr. Herrnstadt acknowledged that some information is collected in the defense sector, but stated that it needs to be made more accessible. There is also a problem with proprietary information. If you are sitting on the worker's side of the table, you have no idea what the key information is because you are not privileged to the details of the contract. The defense industry slammed a congressional proposal to require a "good faith estimate" from defense contractors as to how much work would be kept in the United States. Mr. Herrnstadt argued that this was a very modest proposal. Although the comments made at this symposium have been refreshing, he expressed hope that the industry would review its position as to what type of general information could be released.

The administration must also provide leadership on this issue. There has been a long and fruitful effort on protecting intellectual property. The trade-related problems such as offsets deserve a similar level of attention—not only in aerospace but in other industries as well. Mr. Herrnstadt closed by thanking the National Academy of Sciences for facilitating the dialog on this issue. He noted, however, that action, not just dialog, is needed.

III

PAPERS

Offsets in Commercial and Military Aerospace: An Overview

David C. Mowery
University of California, Berkeley

INTRODUCTION

An offset is a provision in an international export transaction that commits the seller firm to provide technology, to procure locally produced components, or to provide other forms of technical and other assistance to firms in the purchaser nation that go beyond those deemed economically necessary to support the sale. Offsets typically are undertaken by the selling firm in response to formal or informal demands made by the government of the purchasing nation. Although the term has been applied mainly to exports of military systems by U.S. producers, a number of examples exist of similar provisions in U.S. firms' export of civilian products, especially when these products are capital- and technology-intensive goods that are purchased by state-owned or state-controlled firms. Thus, provisions in commercial transactions that closely resemble offsets are common in export markets for telecommunications equipment, power generation equipment, and commercial aircraft and engines. Nevertheless, care must be taken in describing all collaborative activities undertaken by U.S. exporters of these and other commercial products as offsets. As I point out below, a number of strictly economic factors also are driving increased collaboration with foreign firms in many of these industries.

As this last statement points out, the definition of an offset is often very difficult, and (especially in commercial exports) the perception of direct or indirect government pressure, the central defining characteristic of an offset, is often in the eye of the beholder rather than in objective data or other indicators. Another difficult issue in the measurement of the magnitude and economic effects of offsets is the distinction between "direct" and "indirect" offsets. Direct offsets generally are defined to be transactions that are related to the specific product being exported. In the case of a military aircraft, direct offsets may involve commitments by the U.S. exporter to purchase a fixed percentage of the components (typically measured by their value) for the systems being sold to a foreign nation from firms located in that nation. Indirect offsets, by contrast, involve commitments by the exporter to either purchase unrelated products or to provide other forms of technical or commercial assistance to firms in the purchasing nation that are valued at some percentage of the export sale. Needless to say, the measurement and analysis of indirect offsets is considerably more challenging than is true of direct offsets.

Offsets and similar provisions have been important features of U.S. military and civil aerospace exports for much of the postwar period. At least 28 U.S. aircraft and missile systems were manufactured by foreign firms under "co-production" agreements during 1947–1980, and more recent estimates (U.S. Office of Management and Budget, 1990) of offsets as a percentage of U.S. foreign military sales range as high as 98 percent (that is, nearly the entire value of the export sale is being offset by countervailing purchases or transfers). On the commercial side, virtually no new large commercial transport or engine for such aircraft has been developed and manufactured since the 1970s without significant participation by non-U.S. firms in technology development, manufacturing, or marketing.

Offsets resemble other forms of "countertrade," and they are similarly inefficient and trade distorting from an economic perspective. By substituting various forms of barter for monetary transactions, they reduce the efficiency of markets and distort trade flows. But the trade-distorting effects of aerospace offsets are not the primary reason for concern among executive branch and congressional policy makers. Instead, offsets are seen as a cause of potentially welfare-reducing employment effects and technology transfer to foreign firms. U.S. aerospace workers, among the best-paid manufacturing workers in the U.S. economy, are alleged to be losing employment opportunities because of the actions of U.S. firms and foreign governments to "export jobs" in this high-wage industry through offsets. In addition, the transfers of aerospace technologies to foreign firms are asserted to be strengthening foreign competitors to U.S. firms, reducing future U.S. competitiveness and the strength of the U.S. defense industrial base. Finally, the transfers of "dual-use" technologies through military and commercial offsets may be strengthening firms in nations that could be military adversaries in the future, thereby weakening national security.

The issues raised by aerospace offsets are as complex as they are emotional. In this paper I can do little more than survey these issues in a summary fashion, noting the origins and the factors driving the growth of military and commercial offsets. Immediately below, I discuss the factors that have contributed to the growth of international collaboration in military and civilian aerospace. I follow this survey with a summary and assessment of the policy questions created by collaboration in military and civil aerospace. The conclusion suggests some alternatives for consideration by policy makers in addressing the difficult trade-offs and consequences of offsets.

Overall, it is difficult to make a credible case that offsets in both military and commercial aerospace account for any but a small fraction of the sharp declines in aerospace employment since the 1980s. Indeed, the available evidence suggests that indirect offsets now play a more prominent role in military exports, which makes it even more difficult to establish a connection between these provisions and employment losses in U.S. aerospace. Moreover, the effects of offset-related technology transfer on the fortunes of U.S. prime contractors in military and commercial airframes, avionics, or engines are very difficult to identify. Although little or no quantitative evidence has been collected on this issue, anecdotal evidence suggests that the negative consequences of offsets and similar transactions may be greatest among the U.S. firms that supply the prime contractors.

But here and elsewhere in aerospace, the most sensible policies to address offsets are those that are desirable on other grounds. These include continued efforts by U.S. negotiators to reduce the incidence of trade-distorting requirements in foreign government purchases, continued pressure by U.S. negotiators to reduce foreign government subsidization of "national champions," and policies that support adjustment by U.S. workers faced with intensified foreign competition. Rather than developing a specific set of policies to address offsets, the most sensible course of action seems to be one of pursuing liberalization in international trade and investment, while strengthening the weak infrastructure of public support for worker adjustment in this economy.

ORIGINS AND GROWTH OF OFFSETS

In this section I summarize the postwar development of formal and informal offsets in the commercial and military aerospace sectors. As this very short history points out, military aerospace offsets have from their inception been closely linked to broader U.S. political and national security objectives. Indeed, offsets are to a substantial degree a creation of postwar U.S. military export policy. This short narrative also should make clear the complex interdependency between offset-related and other trends in the military and commercial portions of the U.S. aerospace industry. The growth of military aircraft offsets contributed to the subsequent increase in foreign participation in U.S. commercial airframe and engine development and manufacture, and vice versa.

Military Offsets

Co-Production Agreements

During the early postwar period, the U.S. government made extensive use of co-production agreements in weapons sales to allies. Under the terms of such agreements, all or a substantial portion of the assembly of the weapons system purchased by a foreign government was produced by firms in the foreign nation. By guaranteeing that a portion of the costs of a purchase of U.S. weapons would remain within their nation in the form of domestic employment or manufacturing activity, such agreements made these purchases more attractive to foreign governments. Co-production agreements also contributed to the reconstruction of the European and Japanese economies during the 1950s and promoted commonality in the equipment used by Cold War allies. In addition, of course, by expanding markets for U.S. military equipment, co-production agreements lowered the unit costs of weapons systems purchased by the U.S. military.

Between 1947 and 1980, at least 44 different weapons systems, 28 of which were aircraft, missiles, or rotorcraft, were produced by foreign firms in 20 or more countries under licenses granted by U.S. producers. Until the late 1970s, co-production agreements did not involve transfers of systems design data for these weapons. In most cases, licensed co-production began with the assembly by foreign firms of knockdown kits from U.S. producers, followed by expansion in the range of locally produced components for incorporation into a weapons system.

Because they rarely included joint development or design activities, the co-production agreements of the 1947–1980 period transferred little by way of the design and systems integration skills needed for independent entry into the development and manufacture of military or civilian aircraft. But the aerospace industries of such nations as Great Britain, France, Japan, and West Germany retained formidable scientific and technological capabilities—World War II had destroyed production capacity, but had done little to permanently damage the skills of the scientific and technical work force in these nations. As a result, co-production agreements contributed to growth in production capacity in weapons systems and components in both Western Europe and Japan and enabled these industries to more fully utilize their long-established technical capabilities. Their interest in utilizing the technical and production capacity that co-production agreements helped restore, as well as dissatisfaction with the quality and quantity of technology transferred through these agreements, led European and Japanese firms and their governments to demand a greater role in joint development and design of components and systems during the 1970s and 1980s. Congressional passage of the Nunn–Roth–Warner amendment in 1986 created additional opportunities for transatlantic collaboration in weapons development.

These forces contributed to the growth of offset agreements, which in many cases supplemented or replaced the co-production arrangements that had characterized the weapons sales of the early postwar period. Rather than licensed assembly of the weapons system, these offset agreements provided for manufacture (and in some cases, development) by domestic firms in the foreign purchaser nation of components for a portion of the aircraft purchased for domestic use and, in some cases, for a portion of the entire production of the particular aircraft. Thus, in the "sale of the century" in 1976, General Dynamics agreed to assign a major production role for its F-16 fighter aircraft to domestic firms in prospective purchaser nations. This role included the production of components for the aircraft sold to European nations as well as for the aircraft sold to the U.S. Air Force. European producers in Norway, The Netherlands, Belgium, and Denmark were offered 40 percent of the value of production for the aircraft purchased by their respective armed services, as well as 10 percent of the value of production of the aircraft purchased by the U.S. Air Force and 15 percent of the value of production of aircraft purchased by other governments.

For much of the 1970s and 1980s, Defense Department policy treated the negotiation of offsets as an issue to be handled by U.S. military exporters, subject to government approval of any export licenses for sensitive technologies. These laissez-faire policies were altered by the Bush administration's announcement in April 1990 that "certain offsets for military exports were economically inefficient and market distorting" (U.S. Office of Management and Budget, 1990:23–24). The revised policy announced by the Bush administration gave the federal government the right to review offset arrangements negotiated by U.S. military exporters. The new policy led the Defense Department to restrict offsets in the 1990 negotiations over the sale of F/A-18 aircraft to the Republic of Korea: Offsets were limited to 30 percent of the total value of the transaction, not including the estimated value of the production offsets. These restrictions may have influenced the South Korean government's decision to purchase the F-16 rather than the F/A-18.

An even more celebrated episode in the recent history of offsets, of course, is the U.S.–Japanese FSX/F-2 fighter aircraft co-development agreement. This episode reveals the complexities of the domestic politics that now underpin and influence international offset agreements. Japan's Self-Defense Forces have for decades purchased U.S. fighter aircraft, and a long list of these aircraft have been manufactured in Japan under co-production agreements. Just as was the case in Western Europe, however, domestic Japanese aerospace firms began pressuring the Japanese government for a larger role in the development and design of future military aircraft, and in the late 1980s advocated the independent development of an "all-Japanese" fighter aircraft. Japan's defense and foreign ministries resisted this pressure for both economic and political reasons, U.S. government negotiators insisted on a collaborative arrangement between U.S. and Japanese firms, and a compromise was announced in 1987. The next-generation Japanese fighter aircraft, then known as the FSX, would be developed jointly by General Dynamics (based on an extensive modification of the F-16) and powered by a U.S. military aircraft engine.

The announcement of the FSX project created considerable political controversy in the United States. Critics of the agreement argued that it would result in the transfer of critical dual-use technologies to Japanese firms whose capabilities in the commercial aircraft industry would be strengthened. Some also criticized the agreement for its failure to provide for reciprocal access by U.S. firms to jointly developed technologies resulting from the project. Much of this controversy reflected the high pitch of "Japan bashing" in U.S. political debate in 1987–1988, as U.S. firms in many industries seemed to be under severe competitive pressure from Japanese enterprises. But in other respects, the controversy over the FSX reflected the growing political saliency of offset and related technology transfer provisions of U.S. military exports, especially as the Cold War was ending. No longer were these arrangements judged solely in terms of their geopolitical or national security implications. In the post-Cold War era, the consequences of offsets for domestic employment, competitiveness, and economic welfare had for the first time become a topic of congressional debate.

After much discussion, a revised U.S.–Japanese agreement for the development of what became known as the F-2 was signed in 1988. The co-development project has progressed slowly (a prototype F-2 flew in 1995), and estimated development costs now appear closer to $4 billion than the original $1.1 billion estimate of 1987. The entire project, all of which is funded by the Japanese government, is now projected to cost $14 billion, $10.3 billion of which is associated with production—individual F-2 aircraft will cost $80 million (U.S. General Accounting Office, 1997; for earlier estimates of development cost overruns, see U.S. General Accounting Office, 1992). Government officials in both countries view the project as an overly ambitious first attempt at co-development among firms that lack a history of such activities. Despite the considerable political capital invested by U.S. negotiators in gaining access by U.S. firms to "flowbacks" of Japanese-developed technologies that were based on U.S. technical data, according to the U.S. General Accounting Office (1997), few U.S. firms have expressed interest in these technologies. Thus far, experts view the F-2 project as providing modest support for Japanese aircraft firms' system integration capabilities, but there are few examples of commercially relevant technology transfer from U.S. firms to the Japanese participants through this venture.

It is important to note that foreign sellers of weapons systems to the U.S. military services have to meet a number of performance requirements whose effects closely resemble those of offsets. "Buy American" requirements are commonly inserted into appropriations for major weapons systems by Congress, which handicap foreign bidders for contracts within such programs. Purchases by the U.S. military of "large," assembled weapons systems, such as the Harrier aircraft, the Ptarmigan military radio system, or even the Beretta military sidearm, all have contained stringent requirements for final assembly of the products in North American facilities, often by U.S. firms. These provisions are no less trade-distorting than the offset requirements of other nations.

Data on Trends in Military Offsets

Despite (or possibly, because of) their increasing political sensitivity, data on the economic significance of military offsets are very scarce. The U.S. Office of Management and Budget (1990) collected data on offsets during 1980–1987, which are reproduced in Table 1. More recent data on offsets, also included in Table 1, have been published in a recent study by the U.S. Department of Commerce (1996). All of these studies aggregate all military export transactions in reporting the incidence of offsets and therefore do not allow for a separate analysis of the role of offsets in exports of military aerospace products, as opposed to electronics or other weapons systems.

These data support three broad conclusions:

1) They are very sensitive to the inclusion or exclusion of individual transactions. As the 1996 Commerce Department study notes, the exclusion of two large military export sales from its 1993 data on offsets would shift the reported percentage of U.S. military exports accounted for by offsets from 34.4 percent to more than 69 percent.

2) The data for 1980–1987 are similarly sensitive to the inclusion or exclusion of specific transactions. Despite these flaws, a comparison of the data in Table 1 suggests no strong upward trend in the share of U.S. military exports

TABLE 1 Military Exports and Offsets, 1980–1994

Year	Exports (millions of current $)	Offset Obligations (millions of current $)	Offsets as Percentage of Exports
1980	6,541	3,611	55.2
1981	2,507	2,195	87.6
1982	2,594	1,060	40.9
1983	8,703	4,471	51.4
1984	5,523	2,280	41.3
1985	3,860	2,434	63.0
1986	2,209	1,047	47.4
1987	3,037	2,987	98.3
1993	13,945	4,794	34.4
1994	4,792	2,049	42.8

SOURCES: U.S. Office of Management and Budget (1990); U.S. Department of Commerce (1996).

accounted for by offsets. Indeed, if anything, these data suggest a modest downward trend. For the entire 1980–1987 period, offsets averaged 57.2 percent of military exports, whereas for 1993–1994, offsets accounted for 36.5 percent of exports (the Commerce Department study's reported an average of 54.8 percent for the 1993-1994 period that arbitrarily excludes two large military export sales to Taiwan and Saudi Arabia, without any justification).

3) The importance of indirect offsets, which by definition affect exports and employment outside of the aerospace sector, appears to have increased. The Office of Management and Budget study of offsets during 1980–1987 found that direct offsets accounted for 37 percent and indirect offsets 41 percent of all offset agreements (by value). The Commerce Department's 1996 study concluded that, during 1993–1994, direct offsets accounted for 31 percent and indirect offsets 62 percent of all offset agreements by value.

These data are hardly definitive, but they suggest that direct offsets are declining in importance and indicate either a decline, or at a minimum, no evidence of increased offsets in recent U.S. military exports. A 1996 report by the U.S. General Accounting Office of offsets in U.S. military exports to the major purchasers of U.S. weapons systems in Western Europe, the Middle East, and Asia, however, reaches the opposite conclusion, asserting that "Demands for offsets in foreign military procurement have increased in selected countries." The basis for this assertion is unclear—the General Accounting Office (GAO) report relied largely on case studies and interviews. This methodology means that the GAO report's conclusions could reflect pressures and conditions imposed on potential export sales that U.S. firms rejected, rather than the offset content of export sales actually made—the report fails to make a clear distinction on this point.

The GAO report also indicates that indirect offsets are more common in military exports to newly industrializing and developing economies by a comparison with the industrial economies of Western Europe or Canada. This contrast reflects the far more highly developed armaments industries in the industrial economies, which now are anxious for additional work. The NICs and developing economies, however, have smaller armaments industries and (in many cases) broader developmental objectives associated with their offset strategies. But this contrast has two implications: (1) As U.S. arms exports are directed to newly industrialized and industrializing economies, rather than to NATO allies, one might anticipate a continuing rise in indirect offsets, relative to direct offsets, in the near term; and (2) in the longer term, if the newly industrialized and industrializing economies do pursue the development of indigenous armaments industries, their demands for offsets may shift from indirect to direct offsets.

International Collaboration in Civil Aerospace

There is no clear equivalent to offsets in the commercial aerospace industry, although many observers agree that governments often play an important role in demanding concessions as part of their purchases of large commercial transports. The pervasive international collaboration that now characterizes the development, production, and marketing of large commercial transports has a more complex set of causes than government pressure alone. Among these causes, of course, is the legacy of previous co-production and offset agreements associated with foreign purchases of U.S. military aerospace products. And the currently high levels of international collaboration in civil aerospace are likely to produce additional pressures for direct and indirect offsets in future U.S. military exports. In this subsection, I review the development of international collaboration in civil aerospace so as to provide some context for these statements.

The large commercial transport industry consists of two large manufacturing sectors—engines and airframes. With the exception of some large Japanese aerospace firms (Mitsubishi Heavy Industries and Kawasaki Heavy Industries), most major manufacturers of airframes do not engage in engine production, and vice versa. Moreover, since the early 1970s, large commercial transport airframes have been designed to accommodate more than one engine design within a given thrust class. The technological linkages between the engine and airframe industries remain critical—development of a new class of transports typically requires the introduction of a new class of engines—but the economic interdependence of individual engine and airframe manufacturers has if anything declined within the large commercial transport industry during the past 25 years. Accordingly, I discuss trends in international trade and market structure for each of these segments separately and follow this with a brief overview of another important segment of the U.S. industry, suppliers of components and subassemblies.

Airframes

In 1996, U.S. producers of airframes (SIC 3721), which includes large commercial transports and international trade data that are dominated by large commercial transports) exported goods valued at $19.0 billion, a sharp increase from 1995 exports of $13.9 billion and slightly above 1994 exports of $18.81 billion (Table 2). U.S. imports of these products amounted to $3.9 billion in 1996, up by 8 percent from their 1995 level of $3.65 billion (unpublished data, International Trade Administration, U.S. Department of Commerce, 1997). Valued in current dollars, exports display little strong upward or downward trend during the 1989–1996 period, a period characterized by sharp cutbacks in U.S. defense spending and wide fluctuations in orders.

The forces that have led to increased international collaboration in airframe development, production, and marketing (see below) have also reduced the number of firms engaged in overall design, systems integration, final assembly, and marketing of large commercial transports. Since 1985, when Lockheed ceased production of the L-1011, there have been three producers of large commercial

TABLE 2 Aircraft Exports and Imports, 1989–1996 (billions of current year dollars)

	1989	1992	1993	1994	1995	1996
Aircraft exports	14.3	26.3	21.3	18.8	13.9	19.0
Aircraft imports	2.8	3.9	3.7	3.7	3.7	3.9
Engines and parts exports	6.6	6.7	0.2	6.4	6.1	6.8
Engines and parts imports	3.9	5.8	5.2	5.3	4.7	5.6
Aircraft parts and equipment exports	8.7	9.1	9.1	9.4	10.0	11.5
Aircraft parts and equipment imports	2.9	3.1	2.4	2.5	2.5	3.3

airframes: Boeing, McDonnell Douglas, and Airbus Industrie of Western Europe, a consortium of British, French, German, Spanish, and Dutch aerospace firms. The merger of Boeing and McDonnell Douglas has reduced this number to two. Since the mid-1970s, the market share of U.S. airframe producers, whether measured in terms of orders, deliveries, or order backlogs, has declined from more than 80 to 60–70 percent. Much of this decline in U.S. market share was absorbed by McDonnell Douglas.

The importance of military–civil "spillovers," technologies developed for military aircraft and subsequently applied to civilian products, has declined significantly in airframes during the past 40 years. But military markets still account for a substantial portion of the revenues of the major U.S. airframe manufacturers. In 1993, 20 percent of Boeing's products and 60 percent of McDonnell Douglas's products (by value) were sold to the U.S. government, primarily to the military services (U.S. Department of Commerce, 1994). Moreover, for much of the past decade, sales trends in military and civil airframe markets have offset one another—during 1985–1991, defense aerospace shipments declined at an average rate of 2 percent per year, but commercial aerospace shipments grew at an average rate of 11 percent per year.

Table 3 presents the market outlook for airframe manufacturers, based on recent forecasts by the Boeing Commercial Aircraft Company. The data confirm a long-standing trend of slower growth in passenger traffic in the industrial nations of Europe and North America, combined with relatively rapid traffic growth in the developing and newly industrialized nations of Asia and elsewhere (see Table 4). Passenger traffic in the Asia–Pacific region, characterized by rapid rates of economic growth and a relatively underdeveloped intra-regional airline network, is projected to grow more rapidly than traffic in any other region. Within

TABLE 3 Projected Airframe Demand (including replacement), 1997–2006

Region	Number of Aircraft Deliveries (% of total forecast deliveries)
North America	2,460 (33.8)
Asia–Pacific	1,750 (24.0)
Western Europe	2,070 (28.4)
Rest of world	1,000 (13.7)

SOURCE: Boeing Commercial Aircraft Company (1997).

TABLE 4 Projected Annual Growth in Passenger Seat Miles, 1997–2006 (in percent)

Region	Projected Annual Growth of Revenue Passenger Miles
China and Hong Kong	9.6
Commonwealth of Independent States	8.6
South America	7.1
Japan and Korea	6.8
Southeast Asia	6.5
Central America	5.9
Africa	5.5
Southwest Asia	5.4
Europe	5.2
Middle East	4.7
North America	4.2
Oceania	3.9

this region, the domestic air traffic network of China is likely to grow rapidly and (combined with the low level of development of domestic Chinese air transport) may propel China to a position as the largest single market for large commercial transport in the world by the year 2000 (see Dornheim, 1994). The most recent market study by the U.S. Department of Commerce (1994) projects that China will purchase more than $40 billion in new aircraft during the next two decades.

These market projections cover only "large" (80 passengers and more) commercial air transports. Although U.S. firms remain active in the development and production of turbojet- and piston-powered business and general aviation aircraft, they have had little success in small, short- and medium-range commuter (20–40 seats) transports. The U.S. firm that was most active in this segment, Fairchild Aviation, entered into a joint venture with Saab Aircraft in 1980 that introduced the SF-340. The joint venture suffered from technological and management problems, however, and Saab eventually assumed sole ownership of the project. Fairchild has largely exited from developing or producing commuter transports. Boeing acquired DeHavilland of Canada, producer of several popular commuter aircraft in 1986, but sold the firm to Bombardier in 1992.

The absence of U.S. airframe firms from this segment of the industry may have important consequences for their prospects in the developing nations of the world economy that are registering rapid rates of growth in passenger traffic. Short-haul, low-maintenance commuter aircraft are well suited to many of these markets, including substantial portions of China's domestic air transport system. In addition, these aircraft (e.g., the CASA/Nurtanio commuter air transport) are well suited to quick conversion from civilian to military air transport uses, which makes them attractive for purchase by governments seeking dual-use aircraft. In response to this market opportunity, Daewoo of South Korea has been co-developing a 33-seat commuter aircraft in partnership with Dornier of Germany. There are few established U.S. producers of such aircraft with whom Daewoo might team. Nonetheless, most South Korean and other Asian airframe manufacturers appear to see jet aircraft with 100–120 seats as the most promising segment of the market, and this segment is one in which McDonnell Douglas, Boeing, and Airbus Industrie all are able to develop and manufacture products (see Cole, 1994).

Although they are consistent with widespread perceptions within the U.S. aerospace industry that East Asia and the Pacific region are the growth markets of the future, the data in Table 4 also reveal that traffic growth in other developing regions of the world, such as the Middle East and Latin America, is projected to be high for at least the next ten years. Moreover, the combination of large, aging fleets of aircraft and more-stringent noise and pollution regulations mean that the demand for aircraft in the industrial nations will be substantial during this period as well. As Table 3 shows, the largest single market during 1997–2006, measured in units, is the North American market for narrow-body air transports, which will account for almost 34 percent of projected sales of these aircraft. Europe and North America jointly account for more than 60 percent of projected deliveries during this period, significantly outstripping the projected deliveries for the rapidly growing markets of Asia.

All three major airframe producers were active in international alliances during the 1980s and 1990s. Airbus Industrie is itself a complex joint venture of the "national champions" of a number of European governments. Airbus Industrie has thus far not entered into risk-sharing subcontracting relationships or other forms of alliances with non-European firms. The consortium's lack of involvement in such alliances may well be due to the concern by its members, a group that includes a number of state-owned firms, over the employment and associated political consequences of significant shifts of work to foreign alliance partners.

Both Boeing and McDonnell Douglas also pursued international joint ventures during this period. In the case of Boeing, international collaboration emerged out of international subcontracting relationships that began with the fabrication of components for the 747. More recently, Boeing has teamed with a consortium of Japanese aerospace firms (Japan Aircraft Development Corporation, made up of Mitsubishi Heavy Industries, Kawasaki Heavy Industries, and Fuji Heavy Industries). This alliance combines sharing of development costs, limited joint development of portions of the airframe, production of fuselage and other large components, and some joint finance of aircraft sales for the Boeing 767 and the 777. Boeing also has involved Short Brothers of Northern Ireland, Aeritalia of Italy, and Saab-Scania of Sweden in risk-sharing relationships in the development of the 777. But Boeing's most prominent joint venture partner remains the Japanese consortium. The participation by the Japanese consortium in both development projects has been supported in part by grants and low-interest loans from the Japanese government. Similar public financial assistance has been extended to Boeing's European risk-sharing partners.

Boeing has allowed its European and Japanese collaborators, especially the latter, to expand their roles beyond those of traditional suppliers and subcontractors in moving from the 767 to the 777 projects. Having mastered the production technology for the main body sections, wing ribs, and other body parts of the 767, the Japanese are gaining additional experience in the 777 project from their production of fuselage panels, the tail fuselage structure, the aft bulkhead, and the wing center section (the portion of the wing that enters the body of the aircraft). The next step, an equity partnership,[1] could involve foreign partners in designing, developing, testing, producing, marketing, selling, and providing after-sales support for the entire aircraft. Although Boeing now "outsources" a much larger share of the development and component manufacture for these aircraft, the U.S. firm retains undisputed management and marketing control of both the 767 and 777 projects.

McDonnell Douglas adopted a different strategy for international collaboration from that of Boeing. Following a series of unsuccessful attempts at collaboration with European aerospace firms (see Mowery, 1987), McDonnell Douglas largely avoided risk-sharing ventures in new aircraft development. One consequence of the firm's lack of international partners was a lack of new aircraft, as McDonnell Douglas introduced only one new aircraft (the MD-11), which is a derivative of a 25-year-old design, during the past decade. Rather than relying on

[1]Boeing was willing to accept such an arrangement in the 777 venture, but the Japanese participants rejected it, fearing political repercussions in the United States.

foreign risk-sharing partners for the development and manufacture of new commercial aircraft, however, McDonnell Douglas established a venture with the Shanghai Aviation Industrial Corporation, licensing the assembly of MD-80 and MD-90 aircraft for use in China's domestic air transport system. McDonnell Douglas's most recent attempt to enter an alliance aimed at global markets occurred in 1991 with the signature of a Memorandum of Understanding (MoU) between McDonnell Douglas and the state-controlled Taiwan Aerospace Company for the development of the MD-12. The announcement of this MoU sparked considerable political debate within the United States over the "loss" of a strategic U.S. industry and an equally vociferous storm of Taiwanese criticism of an agreement that combined high costs, great risk, and a U.S. partner that had serious managerial, production, and marketing weaknesses. The MD-12 joint venture fell apart within six months of the signature of the original MoU. The failure of McDonnell Douglas to develop a more effective international collaboration strategy contributed to this firm's failure to expand its product line sufficiently to remain a viable competitor and, therefore, was an important factor in the firm's demise as an independent producer of large commercial aircraft.

Aircraft Engines

The trends in international collaboration that characterize the airframe segment of the global aerospace industry are apparent in engines as well. There is no counterpart to Airbus Industrie within the commercial transport engine business, although Rolls Royce was government owned during 1971–1987. Despite the absence of a publicly financed competitor, U.S. producers of commercial jet engines have pursued numerous international joint ventures. Measured in 1992 dollars, U.S. shipments of aircraft engines declined sharply during 1989–1995, from \$25.2 billion to \$16.5 billion. The 1995 figure for shipments, however, represents a modest increase from the 1994 level of \$15.9 billion (1992 dollars), and unpublished Commerce Department data forecast growth to \$17.2 billion and \$22.4 billion, respectively, in 1996 and 1997. Exports have remained steady, valued in current dollars at 6.6 billion in 1989, \$6.1 billion in 1995, and \$6.8 billion in 1996 (see Table 2). Imports, on the other hand, display modest growth during this period, from \$3.9 billion in 1989 to \$5.6 billion in 1996. The leading sources of U.S. imports of these products are France, the United Kingdom, and Canada.

As in the airframe segment, three large firms dominate global production of aircraft engines for large commercial transports: General Electric, the Pratt & Whitney unit of United Technologies, and Rolls Royce of Great Britain. As of 1992, 53 percent of large commercial transports in service in the world's fleet were powered by Pratt & Whitney engines and 27 percent were powered by General Electric engines (U.S. International Trade Commission, 1994); Rolls Royce engines accounted for the majority of the remaining 20 percent. Among these three producers, the firm experiencing the most significant growth in market share during the past two decades has been General Electric, which has made effective use of its joint venture with SNECMA and CFM International.

The market outlook for turbojet and turbofan engines obviously tracks that for airframes closely, and the emerging markets of the newly industrializing regions of East Asia are among the most attractive markets for aircraft engines. The possibilities for "retrofitting" older airframes with new, high-bypass engines mean that some of the industrial nation markets that face significant new restrictions on engine noise may account for a larger portion of future sales. The North American and European markets are likely to account for a majority of near-term demand (i.e., deliveries over the next decade).

The importance of France and Canada as sources of U.S. imports stems from the role of SNECMA as a major source of components for General Electric engines and from the role of Pratt & Whitney of Canada, a large subsidiary of the dominant U.S. and global supplier of engines. Japan and South Korea together account for less than 3 percent of U.S. imports of aircraft engines and parts throughout this period. The prominent role of France as a destination for U.S. exports reflects shipments of General Electric engines to SNECMA for final assembly in France, as well as the large sales of CFM International engines to Airbus Industrie, located in France. In contrast to their minimal role as import sources, Japan and South Korea together account for almost 10 percent of U.S. exports during this period because of the large fleets of long-haul, four-engine commercial transports operated by each nation's airlines.

Despite the absence of a well-subsidized "regional champion" competitor in this industry comparable to Airbus Industrie, all three major global producers of aircraft engines are heavily involved in international joint ventures. Pratt & Whitney and Rolls Royce are teamed in International Aero Engines (IAE), a five-nation, seven-firm venture that involves Mitsubishi Heavy Industries, Kawasaki Heavy Industries, Ishikawajima-Harima Heavy Industries, Fiat, and MTU of Germany, in addition to the two principal firms. In contrast to the Boeing joint ventures with foreign firms, IAE is an entity with considerable autonomy from its "senior parents," Pratt & Whitney and Rolls Royce. IAE's primary product, the V2500 engine, experienced severe difficulties in development that were attributable in part to the efforts of the venture's senior participants (Rolls Royce and Pratt & Whitney) to restrict technology flows among the participants (see Mowery, 1987, 1988). The V2500 engine joint venture has expanded the participation of Germany's MTU and Italy's Fiat beyond their roles in previous projects with Pratt & Whitney in the medium- and high-thrust classes (respectively, 35,000–40,000 pounds of thrust and above 60,000).

The General Electric–SNECMA venture, CFM International, has been active since the early 1970s, developing and manufacturing the successful CFM56, a high-bypass, low-thrust engine that effectively enabled General Electric to reenter the civilian jet engine industry. General Electric also relies on SNECMA extensively as a supplier of components for its high-thrust engines, the CF6 series. Indeed, General Electric has benefited substantially from Airbus Industrie's success, since it is a major supplier of engines for the consortium.

Suppliers of Parts and Components

The industrial structure of the world's commercial aircraft industry is noteworthy for its combination of a very small number of large, global firms engaged in product development, systems integration, final assembly, and marketing, and a very large group of firms, most of which are far smaller and engaged in the production of parts and components for these "prime contractors." Surprisingly little is known about the structure of the U.S. "supplier tier," but estimates of the number of U.S. supplier firms in the 1980s ranged as high as 20,000. Many of these firms supply components for both military and commercial airframes and engines. Since the early 1980s, this population of firms appears to have shrunk considerably, as a result of declining sales of commercial and military aircraft. The U.S. Air Force Association asserted that between 1982 and 1987, nearly 15,000 suppliers exited from the defense aerospace components business (quoted in U.S. Department of Commerce, 1993:20-21). Moreover, the number of U.S. firms supplying Boeing and McDonnell Douglas shrank from more than 11,000 to less than 4,000 during the 1980s, a period of rising shipments of aircraft.

Despite these apparent declines in the number of supplier firms, this sector of the U.S. aircraft industry continues to register a significant trade surplus that if anything has grown since 1989. U.S. exports of aircraft parts and equipment grew from \$8.7 billion in 1989 to \$11.5 billion in 1996, and this sector's trade surplus grew from roughly \$5.9 billion to nearly \$8 billion in the same period (see Table 2). Although the supplier tier appears to have undergone considerable upheaval and considerable exit by firms during the past 15 years, its trade performance since the late 1980s has been quite strong. As I note below, increased international collaboration among airframe and engine producers produces market opportunities, as well as intensified competitive pressure, for this segment of the aerospace industry. Partly in response to the increased international collaboration involving their customer firms, some suppliers also have begun to pursue international joint ventures; a joint venture was formed in 1987 between the Bendix aircraft brake division of Allied Signal Aerospace of the United States and Dunlop Aerospace of Great Britain to supply wheels and carbon brakes to the Airbus A330 and A340 (Donne, 1987).

Motives for International Collaboration in Large Commercial Transports

The central motives for international collaboration in most industries are threefold: (1) access to markets, (2) access to technology, and (3) access to capital. For U.S. commercial aircraft firms, the first and third of these motives have been central to their international partnerships. In the airframe industry, the presence of Airbus has increased the importance of both motives. The second motive has played an important role in the decision of Japanese firms to team with U.S. airframe and engine producers. Moreover, the enhanced technological capabilities of non-U.S. commercial aircraft firms, especially suppliers of parts and subassemblies, has made them more attractive to U.S. firms as potential partners in international joint ventures. Many of the European, Japanese, Taiwanese, and South Korean firms that now seek to enhance their technological capabilities through partnerships with U.S. aircraft firms were aided in their early postwar development by technology-sharing and co-production agreements covering foreign sales of U.S. military aircraft (Mowery, 1987; Moran and Mowery, 1995).

U.S. airframe and engine firms now are more concerned with improving their access to foreign markets and capital because of changes in their domestic and international competitive and technological environment. The technological environment for U.S. airframe and engine producers has changed in several critical aspects. The importance of military research and development and procurement as a source of revenues and technologies applicable to commercial aircraft has declined since the 1950s and 1960s because of growing divergence between military and civilian performance requirements, as well as changes in military procurement objectives (i.e., far fewer large transports and tankers are being developed for military markets). Declining military–commercial technology spillovers and steadily increasing development costs for new products (estimates of the development costs for the Boeing 777 range as high as $4 billion; see Holusha, 1994) have greatly increased the risks associated with new product development. Joint ventures with foreign firms that are able and willing to contribute development funds are one mechanism for spreading these forbidding risks.[2] Moreover, the interest of many foreign governments, including those of Great Britain, Japan, and Italy, in supporting their domestic aerospace industry often means that various public subsidies (e.g., low-interest or no-interest loans) are available to the prospective partners of U.S. firms. For all their protests against foreign governments' subsidies for Airbus, U.S. airframe firms also have reduced their cost of the capital for these large development projects through collaboration with subsidized foreign partners.

The "technology access" motive operates for both U.S. and foreign firms within these joint ventures. International collaboration provides a way for U.S. firms to realize some returns to more-mature components of their corporate technological portfolios, especially in negotiations with foreign firms interested in using an international joint venture as a way to acquire skills in commercial aircraft design and manufacture. Although U.S. firms typically restrict access by their partner firms to specific aspects of a new airframe or engine, they may

[2]According to interviews in industry and government, Boeing decided to enter into a risk-sharing subcontracting agreement with its Japanese partners in developing the 767 only after several potential U.S. subcontractors rejected such a risk-sharing joint venture (Mowery, 1987, 1988).

effectively be able to realize some return to their project management skills and more-mature technological capabilities by bringing a foreign firm into a project as a junior partner or pupil.[3] The nature of these capabilities (they are highly "tacit" and depend on know-how) may make it difficult to realize a profit from them through an arms-length sale or licensing agreement.

It would be a mistake, however, to assume that all of the technology-related benefits in these joint ventures flow from U.S. to foreign firms. The Boeing Company, for example, has benefited from access to high-quality manufacturing skills in its 767 production-sharing arrangements with its Japanese risk-sharing subcontractors. Among the primary foreign technological assets of interest to U.S. firms that sought to collaborate with East Asian firms (e.g., McDonnell Douglas and Taiwan Aerospace) were the capabilities in manufacturing management and quality of these enterprises.

Market access appears to be at least as important as access to capital or technology in the decisions by U.S. firms to seek foreign partners in developing and manufacturing new airframes and engines. As I have noted elsewhere (Mowery, 1994), the increasing importance of nontariff barriers in international trade, including such barriers as performance requirements or offsets, has been responsible for much of the recent growth in international joint ventures in a broad array of industries, from steel to semiconductors and automobiles. The early appearance and enduring importance of international collaboration in the large commercial transport market reflect the fact that it is one of the most highly politicized high-technology markets, perhaps second only to telecommunications equipment. In addition, of course, the combination of rising development costs and slower demand growth in their domestic markets means that rapid and substantial penetration of foreign markets is now indispensable to the success of new products developed by U.S. and European firms.

The commercial aircraft industry's status as an important source of high-wage employment, its dual-use technologies and production facilities, and its frequent requirements for government financial support all mean that foreign governments play a prominent role in their domestic aerospace industries. Within Western Europe, the desire of national governments to sustain domestic aerospace industries that were valued in part for their national security benefits influenced the formation of the Airbus Industrie consortium (see Mowery, 1987). Moreover, many foreign airlines have long been directly government owned or subject to substantial "administrative guidance" in their purchasing decisions. The export markets faced by U.S. producers of commercial aircraft thus often require some form of economic concession to gain access.

---

[3]According to *Aviation Week and Space Technology* (1977:201), Boeing's Japanese risk-sharing subcontractors for the 767 project paid the U.S. firm more than $140 million as a royalty for Boeing's production and design experience, as well as its global sales and product support network.

Building on precedents established in U.S. exports of military aircraft, foreign purchasers of U.S. commercial transports in the late 1960s and 1970s began demanding that their purchases contain some domestic content, perhaps by purchasing components from local producers (in other cases, U.S. airframe and engine firms would guarantee "offsetting" purchases of other commodities, including a large shipment of Polish hams in one celebrated case). During the subsequent 20 years, however, and again in parallel with developments in foreign markets for U.S. military aircraft, foreign governments have demanded that U.S. firms offer more-generous benefits, including opportunities for their firms to participate in developing and producing more-complex components. Needless to say, this form of collaboration involves more-intense interaction and higher levels of technology transfer among the participating firms. The importance of market access as a motive for these international joint ventures is so great that relaxation of any remaining domestic antitrust restrictions on joint product development (in 1993 the National Cooperative Research Act of 1984 was amended to allow joint production ventures) seems unlikely to result in U.S. firms choosing to collaborate with their domestic rivals rather than foreign firms (see Mowery [1987, 1988] for additional discussions of the role of domestic antitrust policy in the international collaborative strategies of U.S. firms).

CONSEQUENCES OF CIVIL AND MILITARY AEROSPACE OFFSETS

Concern over the consequences of civil and military aerospace offsets has increased considerably during the past decade. The brief discussion of the FSX controversy illustrates the changing political attitudes toward military aerospace collaboration. Within the civil aerospace industry, offsets have sparked considerable labor–management friction, and disputes over foreign outsourcing contributed to a lengthy strike against the Boeing Company in 1996.

Criticism of offsets in both civil and military aerospace is based on the belief that international collaboration contributes to job losses in the U.S. aerospace industry and the belief that the technology transfers supported by these agreements work against the long-term competitiveness and potentially the long-term national security of the United States. In this view, U.S. military and civil aerospace contractors, especially the prime contractors, are following in the footsteps of U.S. predecessors in the consumer electronics, semiconductor, and other industries, "giving away the future" through one-way transfers of technology that will ultimately strengthen their competitors and result in the entry by foreign competitors into an industry long dominated by U.S. firms.

On the other side, defenders of collaboration argue that international collaboration yields important commercial and technological benefits to U.S. firms, and that without such arrangements, U.S. military and civil aerospace firms would lose foreign sales, creating more serious employment losses among their U.S. work force. These defenders further argue that the technology outflows are carefully managed and are unlikely by themselves to sufficiently strengthen foreign firms to the point that these offshore partners would become direct competitors of the prime contractors.

It is virtually impossible to evaluate the employment, technological, or competitive effects of international collaboration in a rigorous manner with the available data. Publicly available data on employment and shipments are highly aggregated and cannot be linked to individual firms or product lines so as to allow one to analyze the employment or other effects of individual agreements. Moreover, even if more-detailed data were available, isolating the effects of international collaboration and separating these effects from those produced by numerous other influences (exchange rates, defense spending, business cycles) is impossible. Analyses of aerospace such as those of Scott (1997) cannot separate the effects of offsets from the myriad of other forces affecting aerospace employment or shipments.

Employment Effects

By far the most important factor depressing employment in the U.S. domestic aerospace industry during the past decade is the behavior of defense procurement spending in the United States and other industrial economies. Measured in constant 1997 dollars, U.S. Defense Department spending on procurement has dropped from roughly $370 billion in FY 1987 to less than $240 billion in FY 1997. The presidential budget for FY 1997 requested funds sufficient to purchase 73 military aircraft, a dramatic drop from the requested procurement level of 337 military aircraft in FY 1990 and 497 in FY 1985. Reductions in defense spending also have occurred in most Western European nations, which reduces their demand for U.S. weapons systems and intensifies competition between U.S. and European producers in European and foreign markets. The employment consequences of military and civil offsets are minuscule by comparison with those resulting from these enormous shifts in government procurement.

Two other factors further complicate the analysis of the employment consequences of offsets. First, as noted above, a large and apparently growing fraction of the offsets associated with military export sales are indirect offsets, which involve transactions affecting industries other than aerospace. Obviously, the employment effects of these arrangements are both more diffuse and even more difficult to trace. A more fundamental issue, however, makes simplistic job-counting exercises futile. Foreign purchasers of U.S. exports pay for these products by exporting to the United States—over the long term, U.S. imports and exports must approximately balance one another,[4] and U.S. aerospace exports are indeed

[4]Recent U.S. current-account deficits, reflecting a chronic excess of U.S. imports over exports, have resulted in foreign nationals' accumulating future claims on U.S. output that eventually can be made good by purchasing U.S. exports.

offset in full by U.S. purchases of imported goods. The real effects of offsets are their distortion of the offsetting trade flows—for example, rather than importing Canadian potash in compensation for Canadian purchases of aircraft, an offset means that U.S. citizens increase imports of avionics or aerospace components above what they otherwise would obtain. The employment effects associated with offsets therefore center on the relative labor intensity of the goods imported as a result of the offsets versus those that would have been imported in the absence of these arrangements. The data requirements for the necessary counterfactual model of trade flows are forbidding and prevent a true accounting of the employment effects of offsets. But these effects are likely to be quite small. Indeed, the 1990 study of military offsets by the U.S. Office of Management and Budget, the only recent study to attempt a rigorous analysis of the employment effects of these offsets, concluded that:

> "The effects of offsets on total U.S. employment are minor. That is to say, military sales abroad with contractually required offsets are likely to increase domestic employment by somewhat more (by about 2500 employee years [roughly 600 jobs] per year) than would comparable sales without offsets. This is true largely because offsets are a substitute for (but are less labor intensive than) the imports that would replace them to finance the foreign sales" (1990:53).

The effects of offsets on aerospace industry employment thus appear to be minor. This finding does not justify opposition to public policies designed to aid the adjustment of aerospace workers to the upheavals in their industry resulting from changing patterns of defense procurement. Indeed, this finding underlines the point that a sensible federal policy toward the domestic employment consequences of offsets should be part of a portfolio of federal programs to facilitate adjustment by workers to broader trends of intensified global competition and expanded foreign trade rather than designing adjustment policies that attempt to deal with the specific (and unidentifiable) employment consequences of offsets.

Effects of International Collaboration on Commercial Competitiveness

What are the competitive consequences of collaboration between U.S. and foreign commercial airframe and engine firms? It is difficult to construct a credible counterfactual case to answer this question. The two U.S. firms that have been most successful in establishing and managing these undertakings, Boeing and General Electric, have maintained access to important foreign markets, while simultaneously financing ambitious new product development programs. The U.S. firm that has been least successful in international joint ventures, McDonnell Douglas, was squeezed out of the large commercial transport industry in large part because of its inability to bring new products to market in a timely fashion.

On the other hand, McDonnell Douglas's international joint venture in China significantly enhanced its access to this large market.

Moreover, despite criticism of U.S. firms for "giving away the future" through these joint ventures (see Prestowitz, 1992; Reich, 1986; Reich and Mankin, 1986; and a very different view in Reich, 1990), which critics claimed would build up robust foreign competitors that would enter the large commercial transport industry, there is little evidence of imminent entry by Fiat, Aeritalia, Mitsubishi Heavy Industries, Shorts Brothers, or Saab Aircraft into the large commercial aircraft industry. Indeed, the costs and risks associated with such entry are forbidding and have contributed to the decline in the number of producers of airframes and engines. As the Commerce Department's senior aerospace analyst, Sally Bath, pointed out in her remarks at the National Research Council's June 1997 Workshop on Aerospace Offsets (Bath, 1997), there is virtually no evidence that their engagement in military offset and co-production arrangements significantly enhanced the technological or competitive capabilities of the Airbus Industrie member firms.

The most serious competitive consequences of collaboration between U.S. and foreign airframe and engine firms appear to have been felt by the supplier tier of the U.S. aerospace industry (see Mowery, 1987; Friedman and Samuels, 1992). Many supplier firms are relatively small and lack strong proprietary technological capabilities. In this industry, as in other U.S. manufacturing industries, the relationship between suppliers and the "prime contractor" firms has often been adversarial, with limited sharing of technology, management skills, or financing. The involvement of foreign firms as risk-sharing subcontractors with U.S. prime contractors has strengthened their technological capabilities and has intensified competitive pressures on U.S. supplier firms. These competitive pressures, in combination with reductions in U.S. defense spending, have contributed to the exit of large numbers of firms from this segment of the U.S. commercial and defense aircraft industries. This consequence of international collaboration is worth noting, because it could produce a more restrictive U.S. government policy toward international collaboration in this industry. At the same time that it has intensified competitive pressure on U.S. suppliers, however, increased international teaming in the commercial aircraft industry, which includes the use by foreign firms of U.S. suppliers, has contributed to robust demand for exports of U.S. components and parts.

Outside of the supplier tier within the U.S. commercial aircraft industry, the rise of international joint ventures is much more a response to changing international competitiveness rather than a cause of eroding U.S. competitiveness. In the absence of far-reaching and unlikely changes in the nature of restrictions to market access and significant restrictions on foreign governments' ability to subsidize domestic aerospace firms (an equally unlikely possibility that I discuss below), international collaboration seems likely to continue. Indeed, such collaboration seems to be indispensable to the survival and vitality of the prime contractor firms within the U.S. commercial transport industry.

OUTLOOK

The forecasts of regional demand for large commercial transports discussed above suggest that, despite the importance of the North American market for the next decade, a large share of near-term growth in demand, and a growing share of long-term growth, lies in foreign markets. Moreover, many of these foreign markets, especially those in Asia, are in economies undergoing rapid industrialization and significant improvements in indigenous technological capabilities. As a result, the number of foreign firms with the requisite design, engineering, and production skills to supply U.S. aerospace firms, or to work with them in developing new products, continues to grow. The costs and associated risks of new product development in commercial aircraft also display little sign of abating. Although formal barriers to market access are not growing, nontariff barriers appear to remain significant. Privatization of the airlines that serve as the customers for U.S. exporters of airframes and engines for large commercial transports may reduce the influence of home country governments, but this is a matter of degree. As a result, internationalization of product development and manufacturing activities in the commercial aircraft industry is almost certain to continue.

Two recent developments will affect collaboration in civil aerospace. The first is the rapid growth of the Chinese market for civil aircraft. As China's economy continues to grow rapidly, demand for air travel in China is projected to grow more rapidly than any other market. At present, entry into the Chinese market is closely controlled by the central government, and foreign manufacturers of commercial aircraft face significant demands for direct and indirect offsets. Because overt government pressure for various types of performance requirements in civilian products is subject to discipline under the World Trade Organization's (WTO) Uruguay Round accords, the terms under which China is allowed to join the WTO may constrain these demands for offsets. Successful demands by Chinese negotiators for lengthy transition periods in meeting provisions of the WTO agreement, however, could mean that demands for offsets will remain intense for the next two decades.

A second development with very uncertain consequences for international collaboration in civil aerospace is the merger of Boeing and McDonnell Douglas, which became effective in August 1997. The consequences of this merger for offsets are uncertain for a number of reasons, most prominent among which is the fact that Boeing has committed itself to maintain the product lines of McDonnell Douglas's civil aircraft division for a decade and to manage this division as an independent subsidiary. Beyond this commitment, made to address objections by the Commission of the European Union to the merger, little is yet known about the role that McDonnell Douglas's civil aircraft operations will play within the combined enterprise. To the extent that Boeing is able to control McDonnell Douglas's marketing decisions, it may succeed in removing one (relatively weak) competitor from a portion of its foreign markets. Such a policy could reduce the purely commercial pressures on Boeing to provide opportunities for collaboration with foreign firms to gain access to their markets. Moreover, McDonnell Douglas's substantial operations in China may provide additional opportunities for Boeing to penetrate this important market without providing additional offsets. The merger thus may remove one modest source of pressure on the leading U.S. airframe producer to expand its international collaborative efforts. But the complex nature of the motives for these collaborative undertakings means that the effects of the merger on Boeing's international activities are likely to be small.

The very large aerospace industry of the former Soviet Union has undergone considerable rationalization and restructuring, and at least some Russian firms now appear poised to enter foreign markets, mainly within the Commonwealth of Independent States (CIS), for civilian aircraft. The data in Table 4 forecast growth in revenue passenger miles in the CIS states during 1997–2006 that is second only to China. The CIS market may be relatively unattractive to U.S. or European producers because of severe financial constraints of prospective customers and a consequently high reliance on countertrade. But these markets may prove attractive to Russian aerospace firms, providing a relatively noncompetitive market within which to develop experience, improve quality and product support, expand production, and move down learning curves, etc. Russian aerospace firms are likely to become more effective competitors to U.S. and European firms sometime in the next century, and should it develop, such competition will intensify pressure on U.S. and European firms to provide more generous terms in their export sales. Alternatively, Russian and other CIS aerospace firms may develop into attractive partners for U.S. or European aerospace exporters. The revival of Russia's civilian aerospace sector thus is likely to provide yet another impetus to increased international collaboration.

The revival of Russia's military aerospace industry, combined with the collapse of its domestic market, could also have significant effects on the future of military aerospace offsets. Not only Russia, but virtually all of the NATO governments are reducing their procurement budgets, simultaneously with rapid growth in defense spending in the former Warsaw Pact nations, Latin America, and China. Many of the NATO allies also have significant domestic military aerospace industries, which will seek foreign markets more aggressively in the future. Competition among U.S., European, and former Warsaw Pact aerospace firms in export markets thus seems likely to intensify considerably. Should this forecast prove to be true, pressure on U.S. firms to offer more generous offset terms could intensify. Thus far, U.S. exporters of military weapons appear if anything to be increasing their market share—the Congressional Research Service reports that in 1996, U.S. firms accounted for 35.5 percent of international arms sales, a share more than twice as large as that of the next leading exporter,

Great Britain (Shenon, 1997). Moreover, if the trends identified in the recent Commerce Department survey of military offsets remain valid, this increased market share has not come at the cost of more-generous offset agreements. But in the absence of data on the share of offsets in U.S. military exports in 1996, this conclusion remains speculative.

POLICY ISSUES

Although there is little compelling evidence that offsets have been a significant contributor to recent employment declines in the U.S. aerospace industry, they do distort trade flows and therefore reduce economic efficiency and welfare in the United States. But designing a policy to address offsets alone is ill-advised, in view of their modest economic effects. Instead, dealing with the causes and consequences of aerospace offsets should be addressed as one element of overall policies to deal with international trade and investment, as well as the adjustment needs of U.S. workers affected by these trade and investment flows. In other words, offsets per se are not the central issue for policy in the aerospace industry; instead, U.S. trade policy must focus on reducing governments' resort to subsidies, market access restrictions, and performance requirements in sales of military and civil aircraft. Domestic policies complementing these international policies would provide a stronger infrastructure for assisting workers displaced by trade, technology, and other causes of economic change. Internationalization in the aerospace and other U.S. industries is a fact of life. The policy challenge is adjusting to this reality in international and domestic policies.

Corporate Policies

The discussion in this subsection focuses on government policies. Discussions of offsets and other forms of international collaboration frequently recommend new approaches or responsibilities by managers in consulting with their workers, meeting on an informal or formal basis with policy makers, etc. (see Barber, 1997, and the report of the Committee on Japan of the National Research Council, 1994). Greater consultation and discussion between management and workers within firms engaged in international collaborative ventures seems eminently sensible and advisable. But consultations between management and workers over aerospace firms' plans for future international collaboration are a matter of corporate self-interest, and exhortations from policy makers are likely to have little influence on U.S. firms' use of such discussions. There are no policy-induced impediments to these consultations. Similarly, recommendations for an industry–government advisory council on aerospace issues (National Research Council, 1994) seem unlikely to change either the quality of information available to policy makers or the willingness of corporate managers to share sensitive information on international collaborative arrangements with policy makers. Such a forum might serve as a useful "neutral ground" for union leadership and senior corporate management to discuss long-term trends in the aerospace industry, and such consultations would be beneficial. Beyond this role, however, the utility of an advisory committee whose formation and public statements are not driven by a specific, imminent crisis is very limited.[5] Moreover, because offsets and other forms of international collaboration are affected by a number of policies or impending policy initiatives that are well within the purview of federal policy makers, this discussion focuses on these policies rather than the creation of yet another federal advisory committee.

Trade and Related International Policies

With this in mind, the four most important areas for policy to address offsets and other trade issues in aerospace are (1) completion of the WTO accession agreement with China, ensuring that meaningful disciplines are imposed on Chinese government procurement within the near term (i.e., less than 15 years); (2) continued efforts to strengthen the existing WTO agreements on government procurement, subsidies, and trade-related investment measures (TRIMs); (3) enforcement and improvement of existing U.S.–European Union agreements on aerospace trade, including the 1979 Agreement on Trade in Large Civil Aircraft and the 1992 U.S.–European Union agreement on commercial aerospace subsidies; and (4) completion of negotiations over the proposed NATO Code of Conduct for intra-alliance trade in weapons systems.

The existing array of policy tools with which to address foreign government-mandated offsets is in fact quite extensive, albeit much more comprehensive in its coverage of civil aerospace trade than of military exports. Existing WTO agreements without exception provide extensive loopholes for national security-related transactions, and U.S. trade negotiators have used these in the past to justify a variety of decisions, such as restrictions on machine tool imports. Nevertheless, the proposed NATO Code of Conduct could introduce important disciplines into intra-alliance trade and might serve as a template for eventual extension to cover military sales to other foreign markets for U.S. weapons, such as Latin America and Asia, that are expected to expand rapidly in the next decade.

Two other U.S. policies that exert a powerful influence on the incidence of offsets in foreign military sales also should be reviewed and revised in conjunc-

---

[5]The most credible recent recommendation for such an advisory committee, which was contained in the recent report of the Committee on Japan of the National Research Council (1994), compares the role of an aerospace advisory committee favorably with that of the National Advisory Committee on Semiconductors (NACS). But the (very brief) prominence and influence of NACS were largely driven by the perceived crisis of competitiveness and trade in the U.S. semiconductors of the 1980s, and the NACS' role as a forum for the creation of a long-term "shared vision" has been almost nonexistent.

tion with any multilateral negotiations over military offsets. At present, a substantial exception to U.S. official policy (articulated in a White House April 1990 statement) sanctions foreign governments' use of offsets in export sales financed by U.S. taxpayers through the U.S. Foreign Military Financing (FMF) program. The primary recipients of this aid are Turkey, Greece, Egypt, and Israel, and according to the U.S. General Accounting Office (1994a), all four nations have obtained offsets for FMF sales of U.S. weapons. Imposition of stronger prohibitions on offsets in these sales might at least reduce the extent of U.S. taxpayer subsidies to foreign government use of trade-distorting measures in association with U.S. military exports.

A second set of U.S. policies affecting military offsets that merit review and possibly elimination as part of multilateral negotiations is the "domestic content" provisions that are imposed by Congress or the Pentagon in many procurement programs. The "buy American" provisions and the U.S. Defense Department's insistence on North American sources of supply for weapons being purchased in substantial quantities from foreign producers operate in much the same way as foreign government offsets. Progress in any international negotiations over disciplines on offsets is unlikely without some willingness by the U.S. government to accept greater discipline on its use of these policies.

Domestic Adjustment Policies

Even the most effective set of international agreements, however, will not reverse the powerful trends that are increasing international collaboration in the military and civil aerospace industries. These trends may well increase the instability of aerospace employment and are likely to displace additional workers. Maintaining and liberalizing international trade in goods and technology in aerospace and other industries will remain difficult in the absence of a more coherent program of government assistance to aid workers (as opposed to their employers) in adjusting to the consequences of trade liberalization and economic change.

Current public policies to support investments of public and private funds in work force adjustment are best described as chaotic. Federal policies to support adjustment by displaced workers remain a patchwork of categorical programs, many of which are encumbered with complex eligibility requirements that limit their effectiveness. The U.S. General Accounting Office (1994b) found 154 federal employee training programs with a total budget of $25 billion in the federal budget for 1993–1994, an estimate that includes almost $9 billion in student loan programs. Among the largest programs designed to assist workers' adjustment to economic change are Trade Adjustment Assistance (TAA), which accounted for $215 million in 1993–1994 spending, and the Job Training Partnership Act (JTPA), which (including its Job Corps component) accounted for $5.2 billion.

The TAA program is aimed specifically at the "losers" from internationalization, workers displaced by import competition. Paradoxically, political support for TAA (and, therefore, funding) has crumbled during a period of rapid expansion in import penetration of the U.S. economy. One reason for this program's limited political support may be its poor track record. TAA's requirements for ascertaining that a worker has been displaced by imports, as opposed to the myriad of other potential causes, severely delay the delivery of income support payments and virtually preclude retraining for displaced workers.[6] These problems in this program illustrate the serious handicaps imposed on "adjustment assistance" programs by strict eligibility requirements that are based on the cause of displacement. However politically appealing such requirements may be, they can impair the success of programs designed to support worker adjustment.

The other major federal program for meeting the needs of displaced workers is the JTPA, created in 1982 to replace the Comprehensive Employment Training Act. JTPA's primary assistance to displaced workers, however, remains focused on job search assistance, and the structure of its service delivery system is such that workers with relatively low skills, who are often harder to place in new jobs, tend to be underserved. JTPA also provides relatively little by way of education in basic skills (Cyert and Mowery, 1987), which reinforces a tendency for the program to deal more effectively with relatively better-educated workers within the displaced population.

In summary, the potential economic returns to a stronger set of institutions for the support of investments in the skills of the U.S. labor force appear to be substantial, particularly in an era of rapid internationalization and economic change. But more-effective policies require cooperation among institutions (public education, local government, organized labor, and employers) over which the federal government exercises little direct control and will be difficult to develop in a political atmosphere of considerable mistrust between business and the federal government. Improvements in programs for displaced workers also are hampered by limited understanding of the factors that are most important to program success.

CONCLUSION

For decades, aerospace has presented extreme cases, either in timing or magnitude, of "globalization" trends visible in other industries throughout the U.S. economy. This industry has been closely linked with national security policies, it has been a long-time beneficiary of federal funding of military and civil research and development programs, and it has experienced "alliance-based" internation-

[6]One 1979 study of TAA's operations during the late 1970s found that workers received their first payments on average of 14 months after layoff (Corson et al., 1979). More-recent evaluations of TAA (U.S. General Accounting Office, 1994b) found that program participants received no training in their first 15 weeks of unemployment.

alization to a greater extent than most other high-technology industries in the postwar United States. The issues of offsets and international collaboration are most appropriately viewed in this light. The forces giving rise to them are not unique to aerospace, although they may be more visible and powerful in this sector. And the policies to address their consequences also should not be conceptualized as "unique" in some sense to aerospace. The challenges created by internationalization in civil and military aerospace will be present for decades to come in most U.S. manufacturing and service industries.

As such, it is also worth noting that these challenges are themselves the fruits of a 50-year U.S. policy of support for economic reconstruction and development, support for trade liberalization, and engagement with the international community that has been spectacularly successful. Dealing with these results of success is vastly preferable to dealing with the far more dangerous world that would have attended the failure of such policies.

ACKNOWLEDGMENTS

This paper is based on remarks prepared for the conference organized by the National Research Council's Board on Science, Technology and Economic Policy on "Policy Issues in Aerospace Offsets" held in Washington, D.C., June 9, 1997. Research for this paper was supported by the Alfred P. Sloan Foundation and by the Air Force Office of Scientific Research.

REFERENCES

Aviation Week and Space Technology. 1977. "Japanese doubts rising over F-15, P-3C," June 6, pp. 201–207.

Barber, R. 1997. Assumptions and questions. In C.W. Wessner and A.W. Wolff, eds., *Policy Issues in Aerospace Offsets: Report of a Workshop*. Washington, D.C.: National Academy Press.

Bath, S. 1997. The Airbus experience. In C.W. Wessner and A.W. Wolff, eds., *Policy Issues in Aerospace Offsets: Report of a Workshop*. Washington, D.C.: National Academy Press.

Boeing Commercial Aircraft Company. 1997. *1997 Current Market Outlook*. Seattle, Wash.: Boeing Corporation.

Cole, J. 1994. "Boeing to expand China operations, names new president for unit there." *Wall Street Journal*, August 9, p. A2.

Cyert, R.M., and D.C. Mowery, eds. 1987. *Technology and Employment: Innovation and Growth in the U.S. Economy*. Washington, D.C.: National Academy Press.

Donne, M. 1987. "Dunlop joins U.S. group in Airbus contract bid." *Financial Times*, September 17, p. 6.

Dornheim, M.A. 1994. "Airframe makers temper bullish predictions." *Aviation Week and Space Technology*, March 14, pp. 72–72.

Friedman, D.B., and R.J. Samuels. 1992. How to Succeed without Really Flying: The Japan Aircraft Industry and Japan's Technology Ideology. Paper presented at the National Bureau of Economic Research conference on Japan and the U.S. in Pacific Asia. April 1–3.

Holusha, J. 1994. "Can Boeing's new baby fly financially?" *New York Times*, March 27, Section 3, p. 1.

Moran, T.H., and D.C. Mowery. 1995. Aerospace. *Daedalus*

Mowery, D.C. 1987. *Alliance Politics and Economics: Multinational Joint Ventures in Commercial Aircraft*. Cambridge, Mass.: Ballinger.

Mowery, D.C. 1988. Joint ventures in the U.S. commercial aircraft industry." In D.C. Mowery, ed., *International Collaborative Ventures in U.S. Manufacturing*. Cambridge, Mass.: Ballinger.

Mowery, D.C. 1994. *Science and Technology Policy in Interdependent Economies*. Boston, Mass.: Kluwer.

National Research Council. 1994. *High-Stakes Aviation: U.S.-Japan Technology Linkages in Transport Aircraft*. Committee on Japan. Washington, D.C.: National Academy Press.

Prestowitz, C.V. 1992. The McDonnell Douglas-Taiwan Aerospace agreement: selling off our birthright." *Journal of Policy Analysis and Management* 11:482–486.

Reich, R.B. 1986. "A Faustian bargain with the Japanese." *New York Times*, April 6, Section 3, p. 2.

Reich, R.B. 1990. Who is us? *Harvard Business Review*.

Reich, R.B., and E.D. Mankin. 1986. Joint ventures with Japan give away our future. *Harvard Business Review*.

Scott, R. 1997. Trends and issues in aerospace employment. In C.W. Wessner and A.W. Wolff, eds., *Policy Issues in Aerospace Offsets: Report of a Workshop*. Washington, D.C.: National Academy Press. p. 32-33.

Shenon, P. 1997. "U.S. increases its lead in world market for weapons." *New York Times*, August 16, p. A3.

U.S. Department of Commerce, International Trade Administration. 1993. *U.S. Industrial Outlook 1993*. Washington, D.C.: U.S. Department of Commerce.

U.S. Department of Commerce, International Trade Administration. 1994. *U.S. Industrial Outlook 1994*. Washington, D.C.: U.S. Department of Commerce.

U.S. Department of Commerce, Bureau of Export Administration. 1996. *Offsets in Defense Trade*. Washington, D.C.: U.S. Department of Commerce.

U.S. Department of Commerce. 1997. Unpublished data, International Trade Administration, U.S. Department of Commerce, Washington, D.C.

U.S. General Accounting Office. 1992. *U.S.-Japan Codevelopment: Update of the FSX Program*. Washington, D.C.: U.S. General Accounting Office.

U.S. General Accounting Office. 1994a. *Military Exports: Concerns over Offsets Generated with U.S. Foreign Military Financing Program Funds*. Washington, D.C.: U.S. General Accounting Office.

U.S. General Accounting Office. 1994b. *Multiple Employment Training Programs: Conflicting Requirements Hamper Delivery of Services*. Washington, D.C.: U.S. General Accounting Office.

U.S. General Accounting Office. 1996. *Military Exports: Offset Demands Continue to Grow*. Washington, D.C.: U.S. General Accounting Office.

U.S. General Accounting Office. 1997. *U.S.-Japan Aircraft: Agreement on F-2 Production*. Washington, D.C.: U.S. General Accounting Office.

U.S. International Trade Commission. 1994. *Industry & Trade Summary: Aircraft and Reaction Engines, Other Gas Turbines, and Parts*. Washington, D.C.: U.S. International Trade Commission.

U.S. Office of Management and Budget. 1990. *Study of Military Offsets*. Washington, D.C.: U.S. Office of Management and Budget.

The Policy Context for Military Aerospace Offsets

Kenneth Flamm
Brookings Institution

With the end of the Cold War, military establishments around the world have been decreasing their force structures and spending by significant amounts. This decline in defense spending has made the linkages between the economics of the maintenance of national defense establishments and political–military security issues clearly visible as never before. One issue that is much debated, and highlights the linkages between the economics of international trade in armaments and U.S. national security interests, is the question of American policy toward so-called "offsets."

In this paper I examine the larger context in which the offsets issue is embedded. I argue that offsets are just one dimension—and not necessarily the most important one—of a much larger issue facing U.S. policy makers. Put most starkly, the United States might choose to vigorously promote its armaments exports through active export promotion policies (that match or even exceed what is done by other exporting nations) so as to further strengthen its defense industrial base and lower the acquisition costs of its defense systems. The United States can do this without paying a great deal of attention to the increasingly competitive atmosphere for exports of advanced military capabilities that it will be reinforcing. Or, the United States might instead try to work with its allies' competitive armaments industries to work out some regime that restrains at least some elements in international competition for sales of advanced weapons systems so as to reduce the proliferation of the most advanced capabilities and reduce the urgency (and resource requirements) of development programs for new and ever more sophisticated systems. The United States, in many respects, supplies resources used by both itself and its allies in fielding the most advanced systems competing in the international marketplace. Offsets are both an element in arms deals through which American firms compete against others in global markets and an instrument through which the United States ultimately ends up supporting both U.S. companies and their current and future international competitors.

From a policy perspective, there are two big questions surrounding the use of offsets in defense trade. First, should the U.S. government attempt to unilaterally "countervail" foreign government policies (such as offset requirements) designed to improve the bargaining power of their firms against American suppliers? Or should the United States instead (or in addition) work toward a common international "rules of the game" for competition among potential suppliers of advanced military technology that might both create a "level playing field" for competition and reduce economic pressures to sell ever more advanced technology to ever more dubious customers? Second, how can taxpayer and national interests in maximizing returns on defense technology investments be better aligned with private returns from technology transfer in situations in which the economic impact of the technology transfer may be felt well beyond the boundaries of the firm negotiating the transfer?

In this paper I first examine the current economic environment for defense industries around the world. I then characterize the "big picture" that is stimulating more intensive competition among the major defense suppliers selling the most advanced systems into international markets, and I identify the major regions competing in this global market. Next, I discuss the government policy instruments affecting this competition and analyze the particular role of offsets and offset policy as a dimension of international competition. In the final section I identify policy issues tied to the use of offsets in global defense trade.

THE CURRENT ECONOMICS OF THE DEFENSE INDUSTRY

A key economic fact is that the cost structure in many key high-technology defense industries is dominated by various flavors of economies of scale—in assembling and sustaining essential design capabilities, in systems research and development (R&D), in start-up costs, in production capacity, and in learning curves. The price of entry into development and production of the most advanced weapons systems is a large fixed investment, with unit costs declining sharply as the scale of production increases.

A fundamental element of the national security policy of many nations (including most U.S. allies) is the creation and maintenance of their own independent, autonomous capabilities to produce at least some advanced weapons systems. During the 40 or so years of the Cold War, defense spending was large enough in most countries with pretensions to producing advanced weapons to enable production of these systems in sufficient volumes to at least approach affordability. With the widespread decline in national defense budgets, however, the only way in which many nations will be able to maintain a viable industry is by exporting a much larger portion of their output to overseas customers. This is true in Western Europe, where despite trans-European defense industrial consolidation and halting steps toward a single European defense market, tremendous economic pressures to export leading-edge systems outside the North Atlantic Treaty Organization (NATO) alliance remain in force. It is equally true in Japan, where, with the active support of the Ministry of International Trade and Industry (MITI), Japan's defense industry is currently mounting a campaign to relax current policies prohibiting defense exports. It is even true in the United States, where, since 1995, formal conventional arms transfer policy for the first time has explicitly recognized the economic impact on the domestic industrial base as a considerable factor in decisions on arms exports.

THE BIG PICTURE

In even the medium run, lessened inhibitions on the export of advanced weapons—and increased competition for these sales among the United States and its allies—may have significant impacts on the political and military balance in many regions. In the long run, because retention of a significant technological advantage over adversaries is critical to U.S. military strategy, proliferation of advanced capabilities through the export of weapons by U.S. allies may ultimately be the threat forcing the United States to once again increase its own defense spending and accelerate the development of new generations of systems at a time when budget realities allow little margin for doing so without sacrificing other national priorities.

One excellent example of this phenomenon is use of the so-called "gray threat" (as a recent RAND Corporation study described it) to justify the rapid development of the F-22 fighter. With the imminent production of European fighters (e.g., Eurofighter, Rafale, Gripen) that are beginning to approach the quality of current U.S. front-line fighters and the necessity for the Europeans to export these aircraft to reduce their unit cost, it is likely that U.S. forces will need to deploy even more-advanced fighters in the not too distant future so as to guarantee the assumed substantial margin of superiority over aircraft in the hands of conceivable adversaries. Indeed, once it seems likely that U.S. allies will be willing to sell a relatively potent system to a foreign buyer, there is a considerable argument for instead supporting the sale of an equivalent U.S. system on the grounds that the United States might as well reap the political and economic benefit—and the advantages of a closer military relationship—for itself. In effect, given sufficient competition from U.S. allies, there is a perverse but compelling logic for the United States becoming its own "gray threat."

Thus, there is a complex, self-reinforcing dynamic at work. With declining defense spending, exports have become critical to the very survival of most defense industries outside the United States. Retention (or creation in some cases) of economically viable, indigenous defense systems capabilities is viewed as fundamental to national security in many nations, leading to aggressive economic competition for defense export opportunities. The increasing economic pressure to export ever-more-advanced capabilities, in turn, may alter delicate strategic balances in sensitive regions. Changes in the strategic balance may trigger even greater or wider interest in acquiring advanced systems, and ultimately, create more pressure to accelerate the pace of development of new systems by the most advanced military powers.

One can dimly imagine two possible new equilibria: a regime with much higher levels of defense spending in which the economic pressure to export the most advanced capabilities has subsided to more manageable levels, or, alternatively, the construction of a more cooperative regime for arms sales in which a handful of military powers with any realistic potential to develop the most advanced military systems agree to some degree of mutual restraint on exports to third parties, perhaps in exchange for some program of industrial and technological cooperation that assures the survival of core defense industrial capabilities deemed essential to national security. This last idea has gone by various names—a suppliers' cartel, an "inner circle," etc.—and is probably best viewed as an experiment to be pursued rather than a crystal clear vision of a particular endpoint.

The current market for defense exports provides excellent illustrations of the fundamental linkages and tensions between U.S. security policies and U.S. policies that promote defense exports. In the discussion below I focus on process and organizational issues as well as the particular policies that are pursued.

THE COMPETITORS

To begin, there are basically three countries that sell advanced weapons systems worldwide: the United States, Europe (which in practical industrial terms is beginning to move toward a single European conglomerate in some, though not all, defense sectors), and Russia. Japan has capabilities in defense systems that are highly advanced, but at least until now, has enforced a self-imposed ban on exports. China does not produce the most sophisticated systems, but is an important exporter of middle- and low-end equipment.

U.S. industry is the 600-pound gorilla, accounting for about one-half of worldwide arms transfer deliveries.[1] At least one reason for this is very simple: The United States spends far more on developing new technology and systems. Figure 1 shows the distribution of defense R&D spending between the United States, its NATO allies, and Japan in 1994. The United States accounted for over 70 percent of the total.

The United States enjoys a similar advantage in the size of its internal market

[1]Arms Control and Disarmament Agency figures show almost exactly a 50 percent U.S. share of arms deliveries over the 1992—1994 period.

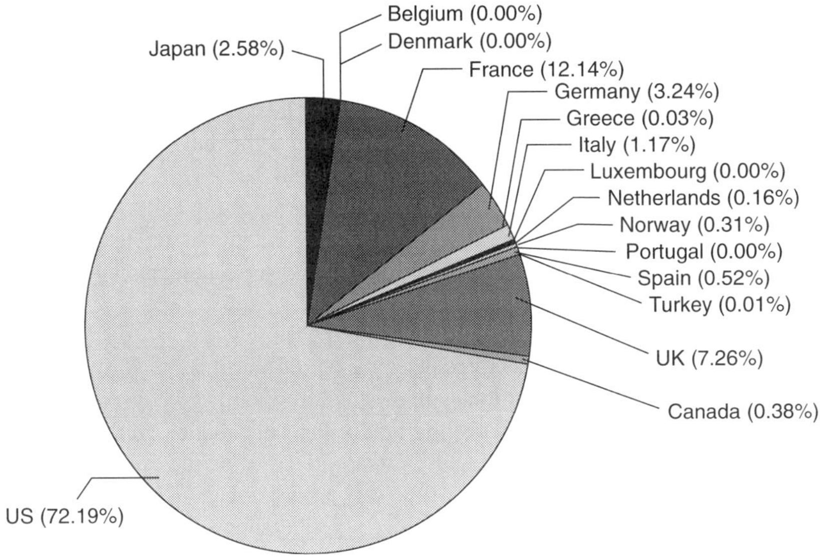

FIGURE 1 Distribution of alliance R&D.

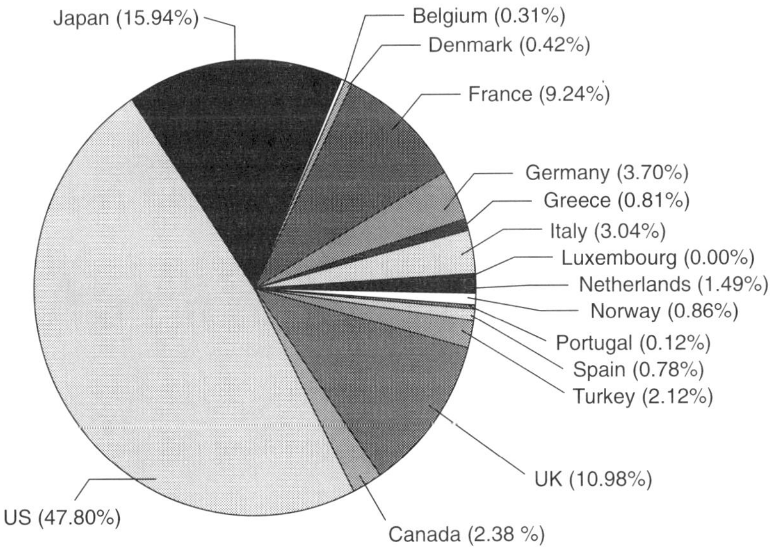

FIGURE 2 Distribution of alliance procurement.

(i.e., its defense procurement budget). Figure 2 compares U.S. procurement with the remainder of its NATO allies and Japan. The United States was slightly less than half of the total; the runner-up was Japan, with about 16 percent of the pie. (This suggests, incidentally, that if Japan were ever to seriously enter the global armaments markets, its domestic output makes it an immediate contender to be the second most efficient producer of military systems.)

Indeed, the real question is—with over 70 percent of allied R&D, how is it that the United States has only a 50 percent share of the global defense trade? Is the United States that inefficient? Are the performance advantages of American systems that much more costly at the margin?

Casual observation suggests that the United States is not grossly less efficient than its allies, and whereas the squeezing out of marginal performance advantages on the bleeding edge of the technological frontier may be disproportionately costly, this too seems unlikely to explain the bulk of this gap. Rather, it seems likely that the United States, through a variety of policy choices, has in effect subsidized the development of high-technology weapons systems by its closest allies. The mechanisms have included a deliberate policy of liberal transfer of technology, on the cheap, to allies through co-production, licensed production, and co-development programs and a variety of policies (such as waiver of recoupment of R&D charges on export sales of components and systems, intellectual property policies, etc.) that make it possible for foreign competitors to acquire some of the key components of high-technology weapons systems at prices that may approach their marginal cost of production.

Buying a U.S.-built radar design at only a modest premium over production cost and inserting it in a European fighter, for example, allows the European systems integrator to market a state-of-the-art platform without investing in an enormously costly development effort on the radar subsystem. Having an American defense contractor work as a joint venture partner on air-to-air missiles may provide access to technologies developed at great U.S. taxpayer expense for a much more modest cost.

This is not to say that doing this was irrational from the U.S. perspective. Strengthening U.S. allies militarily (including their industrial capabilities) was a security interest of the United States that was given priority over possible implications for longer-term economic competition during the Cold War. Often, U.S. allies built protective walls around their defense markets, and giving them access to U.S. technology was one of the prices for slipping over the walls. The decision was an economic one—selling them something, with some return, was better than selling them nothing, and earning no return on technology that in any event had already been paid for. The decision reflected a political judgment—industrial cooperation strengthened U.S. alliances. And the decision was a military one—given that the United States would fight alongside its allies, why not give them the same equipment to use and build a greater operational military coherence?

The structure of incentives within the U.S. acquisition system was another factor promoting bargain basement technology transfer to U.S. allies. U.S. defense contractors were, after all, contractors. The costs of technology development were funded primarily by the taxpayer (although the property rights to the technology, when not used by the U.S. government, generally went to the contractor). Unlike the situation in commercial high technology, a company did not have to define a pricing structure for its output that allowed for a reasonable return on technology investments to be recovered so as to remain viable. Furthermore, because often there were competing U.S. contractors able to offer competitive solutions, foreign governments—with considerable monopoly power—were able to play the contractors off against one another so as to negotiate the most favorable possible terms in acquiring U.S. technology. U.S. policy, because the government was forbidden from favoring one contractor over another in competing for foreign sales, did nothing to improve the bargaining position of U.S. firms and the increased economic rent earned on taxpayer-funded military technology investments.

U.S. contractors, of course, always had their own economic self-interest to guide their decision making. If a company decided, for example, to transfer technology representing a taxpayer investment of $4 billion to Japan for $800 million in licensing fees, it presumably was making the judgment that, in the long run, its potential returns on sales lost to future Japanese competition making use of those technologies was valued at less than $800 million. But if government investments in similar technologies were also earning returns for other U.S. companies, it is easy to see how the company's calculation of a floor on what it would be willing to accept for use of the technology, and a national calculation, might logically diverge. I return to this point below.

In short, both the data and the structure of the U.S. acquisition system naturally lend themselves to speculation that what Figures 1 and 2 really depict is the United States shouldering much of the burden of development cost for systems procured and built by its allies. That is, U.S. policy, in addition to underwriting the cost of sustaining the most formidable and effective defense industry in the world—its own—also in effect underwrote its own industry's principal competitors. U.S. policies supporting defense exports are at least a part of this story.

GOVERNMENT POLICY

U.S. policy supports defense exports through three principal avenues:

1. Granting of export licenses. Weapons systems and major system components are all subject to export control. In principle, licenses are only granted when it is in the security interest of the United States; an explicit recognition of the role of arms exports in strengthening the U.S. industrial base was added by the 1995 Clinton administration conventional arms transfer policy. There are no broad criteria or principles that guide decision making on license applications—the meth-

odology is explicitly case by case, with a regional focus and with no guarantee of logical consistency within or across regions. The concept of "benchmarking"—producing general guidelines detailing under what circumstances differing levels of advanced technology could be released as a tool to improve the consistency and coherence of the licensing process—has been the subject of some discussions, but never broadly implemented.

In practice, the de facto process by which licenses are granted boils down to a nation requesting a license to learn about or actually buy something (which not infrequently follows informal contracts with a U.S. contractor wishing to sell something), followed by an interagency review process in which the U.S. departments of Defense (DoD), State, Commerce, Energy, the Arms Control and Disarmament Agency, the intelligence community, and possibly the National Security Council can play significant roles. The agencies not infrequently have different views—economic and trade interests versus security considerations versus proliferation concerns versus diplomatic issues—and as the Nolan Commission recently observed, "Bureaucratic warfare rather than analysis, tends to be the *modus operandi* in what is often a protracted process of plea bargaining and political compromise that may not reflect long-term national objectives." The administration of the system is also very far from space age. Needless to say, the significant potential for uncertainty and delay built into this process—albeit now much improved from a business perspective—can remain an obstacle to exports.

Congressional prohibitions have placed further restrictions on policy makers in specific cases of regional arms transfers. On the other hand, one can argue that, with the increased recognition of economic benefit as a legitimate arms export policy objective, the system has been gradually tipping toward a presumption that—excepting particularly disreputable would-be customers—if someone else is able and willing to sell a particular capability to a buyer, then it might as well be the United States.

2. Diplomatic and administrative support. As the U.S. foreign diplomatic infrastructure became aware that encouraging U.S. exports was a priority for current government policy, a greater involvement in even-handed support to U.S. contractors in winning competitions for military exports developed over the past few years. The support took the form of sharing unclassified insights on what is going on within often opaque budget planning in foreign governments, embassy lobbying with local government officials, U.S. military lobbying with foreign militaries, and senior political appointees lobbying hard for U.S. solutions with their foreign counterparts. In my experience, this has been perhaps the most important and effective element of U.S. policy support for military systems exports.

On the other hand, I have also observed questionable excesses. One example—an East Asian one—was that of an ambitious young ambassador, who with minimal interaction with local U.S. military staff (but presumably greater contact with the U.S. contractor eager to make the sale), was pressing local defense officials hard to buy an advanced military helicopter. Behind the scenes, senior military staff from the DoD's Pacific Command were scratching their heads in befuddlement, observing that the local military had not yet absorbed the tons of recently acquired equipment they were already struggling to master. Furthermore, what was the country's neighbor—also a U.S. ally—going to think of this proposal? Ultimately, once again the calculus ended up boiling down to the fact that if some enterprising salesman—official or unofficial—convinced the locals that they wanted something, then it might as well be the U.S. that does the selling.

3. Financial subsidies to exports. U.S. defense contractors have lobbied successfully for some new financial supports for defense exports by arguing for a "level playing field."[2] The two principal subsidy proposals that have attracted attention in recent years are a waiver of R&D recoupment charges on exports sales and an export loan guarantee facility similar to Exim Bank programs for non-defense exports. The leveling argument has both a domestic component—armaments should receive the same type of treatment that other goods receive—and a foreign component—foreign governments give their firms financial support in exporting, and therefore the United States should too.

This logic is attractive at first glance, but it has two problems. The first one is the implicit assumption that defense exports are—putting aside the special nature of their customers and application—like other traded goods. The second problem is the assumption that broadly focused export subsidies are likely to be a cost-effective tool for increasing export sales.

Generally, weapons systems are *not* like other traded goods in that a national security exemption has exempted them from the subsidy and antidumping disciplines of the General Agreement on Tariffs and Trade (GATT). Thus, although it is true that producers of industrial goods making use of R&D that is funded by other government agencies are not forced to pay a R&D recoupment charge (charges to foreign customers covering a portion of the government's investment in R&D), the extent to which those export sales of goods can be subsidized by government are severely limited by the ability of foreign competitors to seek countervailing duties and antidumping orders. No such restraints apply to weapons systems, which are presumed to be covered by the national security exemptions in the GATT. In fact, one could reasonably argue that "dumping" (pricing exports below full average cost of development and production) is normal practice in international competition in defense systems, unlike virtually any other class of traded goods facing the disciplines of the GATT.

The waiver of R&D recoupment charges (charges to foreign customers covering a portion of DoD's development costs for a system) has some particularly important economic implications in calculating the economic benefit to DoD from

[2]Another form of support for DoD exports that periodically has been the subject of discussion is DoD payment of the costs for its participation in air shows and trade expositions where weapons systems are exhibited to potential customers. Current policy is to pay for such participation in the most important such shows.

defense exports. First, it means that the benefits will be felt mainly through cost declines derived from production-scale economies and learning curves (and possibly through avoidance of shut-down and start-up costs when exports keep lines "warm"), and not through spreading of R&D costs over a larger output. Second, as already stated, it means that foreign users of defense components have potential access to U.S. technology at marginal cost, enabling them to be competitive in systems where they might otherwise be unable to compete against U.S. producers.

R&D recoupment charges have been waived since 1992 for commercial sales. For foreign military sales made on a government-to-government basis, DoD has long had the discretion to waive R&D recoupment charges on sales to NATO, Japan, Australia, and New Zealand and has routinely done so. Since 1996, congressional authorization to do so in other cases has existed.

In 1996, Congress also authorized a system of loan guarantees for defense exports that could support up to $15 billion in sales. This system is required to be "self-financed" through fees charged to the buyers (cynics argue that the Office of Management and Budget "scoring" system used to assess the probability of default in fixing these fees introduces some element of subsidy vis-à-vis market rates, whereas others argue that the paperwork required by the system should be counted as a cost). To date, after about a year of operation, not a single export sale has actually made use of this self-financed program.

Doubts about the efficiency of general subsidies as a tool to promote defense exports are raised by an analysis of actual markets for defense systems. DoD's 1994 forecast of arms exports[3] divided arms deliveries into two categories—goods already under contract for future delivery and products not yet under contract. The global split for worldwide arms trade for 1994–2000 was about 50–50 in these two categories.

Within the "not yet under contract" category, foreign purchases were divided into three categories: (1) the U.S. was the only source for the system the customer was likely to specify, (2) the U.S. was not in competition (it did not produce a cost-effective or competitive system, or the United States did not sell to a particular customer as a matter of foreign policy), and (3) the United States was in competition against other foreign arms producers. Of deliveries during 1994–2000, only 11 percent were in this last category (see Figure 3). The inference was that U.S. arms exports would be at least 48 percent of world sales over this period, and at most 59 percent. Therefore, absent any policy changes, no more than about 10 percent of a 50–60 percent market share was in play when changes in export policy are being discussed (see Figure 4). The conclusion was that an

---

[3]Requirements-based forecasts by the intelligence community were aggregated into an unclassified format and published as *Worldwide Conventional Arms Trade (1994—2000), A Forecast and Analysis,* Office of the Undersecretary of Defense (Acquisition and Technology), U.S. Department of Defense, December 1994.

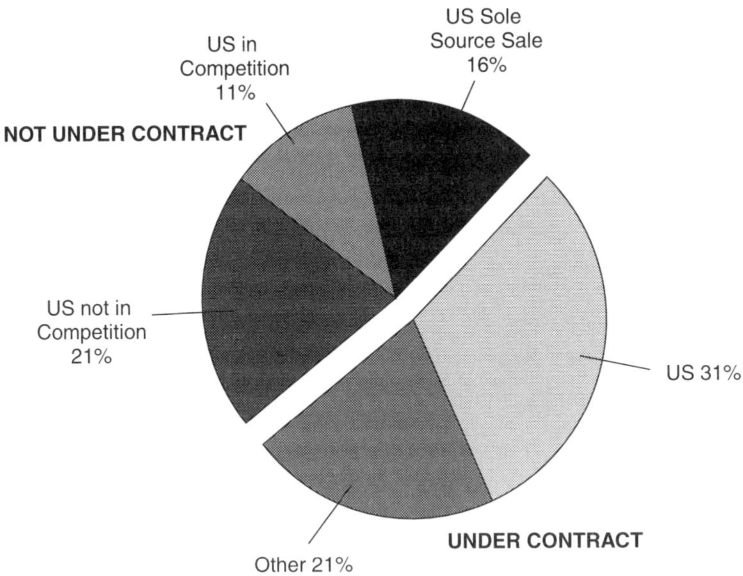

FIGURE 3 Supplier market share of total worldwide arms deliveries (1994-2000).

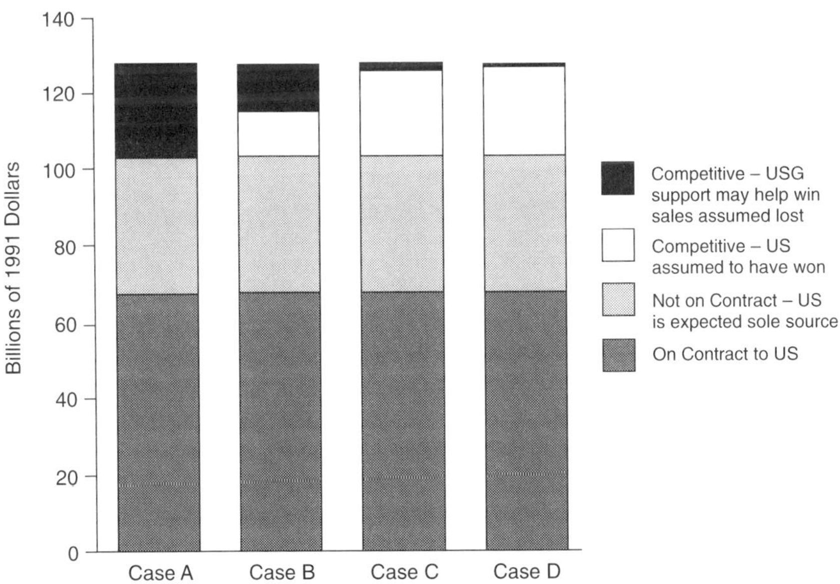

FIGURE 4 Bounds on U.S. arms exports (1994-2000).

efficient export subsidy policy should be selective—picking customers and sectors in which real competition was in evidence and in which it was likely to have a significant impact on DoD's industrial base.

OFFSET POLICY?

One policy instrument that did not make the list above is U.S. offsets policy. This is because, officially, the United States has no offsets policy. To understand why, first I define exactly what an offset is.

Most simply put, an offset is a transfer of goods, services, or other commitments by a vendor and customer that are bundled with the sale of a good or service, but are not required to effectively support the original sale. Offsets may typically include such things as transfers of technology, agreements by the seller to purchase from local suppliers with some connection to the buyer, agreements to invest in production or other facilities in geographical proximity to the buyers, or agreements by the seller to meet certain performance targets (e.g., export requirements) or undertake other unrelated activities (e.g., countertrade) on behalf of the buyer. One can even point to certain transactions that might be regarded as "reverse offsets," with the vendor reducing the price or providing additional services in exchange for commitments by the buyer that would not normally be part of a "straight" sale (e.g., agreement by aircraft vendors to reduce prices in exchange for buyer agreements to exclusively purchase their product over some future period).

Offsets, if defined merely as the activities listed above, are not uncommon in purely private arrangements between private companies operating in today's global markets. Many agreements associated with the rapid growth of so-called "strategic alliances" among multinational companies tie other commitments and activities in with the sale of goods and services to strategic partners. This "barter" of rights to technologies, or other goods and services, transferred between companies, tied to an agreement in which cash may also flow between companies, is common among high-technology companies. It might be argued that this is a reasonable way of dealing with the high degrees of risk and uncertainty in valuation of the future economic impacts of technology.

It is therefore not difficult to see why there is no official U.S. offsets policy. Officially, offsets are a purely private matter between two companies, and there is no reason for the government to intervene. In fact, if it did intervene officially, the government might find itself enmeshed in the details of a competition that might involve two different American firms. Although government officials would have no problem supporting a proposal by an American firm over foreign competitors, favoring one U.S. company's proposal over another's is absolutely forbidden.

But at another level, the U.S. government vaguely disapproves of offsets, which it sees as a distortion introduced into international trade that it would prefer to end. What really makes an offset a matter for official government displeasure is the overt or covert intervention by another government into the terms of what otherwise might be a purely private transaction.

At a still deeper level, the U.S. government is a practical and enthusiastic supporter of offsets used to promote defense sales. If there are going to be offsets, goes the logic, the United States should be able to provide offset packages that will allow the United States to win in competitions with others. Pragmatic U.S. defense officials in recent years have often lobbied for U.S. solutions in foreign defense markets (for both economic and national security reasons), and part of that lobbying has been to assure foreign customers that U.S. suppliers would be allowed to provide "competitive" offset packages.

This highlights the fact that at the deepest level, there is a de facto offset policy when the offsets involve (as they usually do) technology transfer. Such military sales (and many commercial aerospace sales that bundle an offset involving sensitive dual-use technology transfer to a foreign producer) must be approved through the export control process. Approval in this process typically has (at least in recent years) considered the availability of similar technology through offsets provided by foreign firms competing against the American companies seeking export licenses. Thus, the official policy leverage over privately negotiated offset packages has been, and continues to be, approval of export licenses.

ISSUES FOR OFFSET POLICY

Government-mandated offsets in military aerospace connect three economic issues. First, foreign offset requirements can be used to affect the terms of private transactions in ways that improve deals from the perspective of foreign purchasers of U.S. products and technologies, or further other foreign government objectives. Second, because offset requirements in defense often involve the transfer of U.S. technologies funded by public budgets, the incentives of a firm selling such a technology to a foreign buyer can diverge from a more inclusive national calculation of the costs and benefits of a particular deal. Third, offsets can be viewed as a potential backdoor around disciplines imposed on trade and investment by the GATT and the World Trade Organization (WTO).

Bargaining Power and the Terms of Transactions

Offset requirements are imposed by foreign governments as a bargaining tool to influence the terms of transactions involving export sales of American defense technology and products. One objective may be to increase the bargaining power of domestic firms in negotiations with U.S. defense suppliers. By pairing a national preference in defense procurement with government-mandated offset requirements on imports of foreign products or technologies, governments can improve the terms of deals struck between local firms and American partners. Where local firms might otherwise be competing against one another by offering ever more favorable deals to potential American partners, offset requirements can set a minimum floor on the types of deals with local firms that are acceptable to the government, increasing the bargaining power of local firms vis-à-vis possible American vendors. A minimum set of requirements can be designed to improve the terms on which foreign goods and services are purchased by local firms. Obviously, in "commodity" markets with many vendors and a price that approximates long-run costs of production, there is little scope for such policies to accomplish much. In imperfectly competitive markets—such as defense—with small numbers of sellers or buyers, however, such policies can significantly effect the terms on which deals are cut.

Government-mandated offset requirements may also serve other political and economic agendas:

- In economies where government has an explicitly developmentalist view of its role in promoting industrial growth, offsets can be an element of a national industrial policy.
- In economies where government has a major influence on the behavior of certain non-defense sectors (because of public ownership or regulation), governments are frequently tempted to impose formal or informal offset requirements on procurement from abroad that are linked to politically popular (though economically debatable) goals such as jobs or export creation. The same objectives can motivate offsets in defense as well.
- In all countries, defense purchases (closely linked to the aerospace sector) are undertaken by a single customer (the government) with a non-economic goal (the national security). Transactions involving domestic and foreign defense firms (and non-defense goods and services with defense applications) are scrutinized and shaped by all governments to reflect their perceived national security interests. The latter may have an explicitly economic component (protecting or stimulating the defense industrial base).

The policy issue for the United States is that it is apparently acceptable for a foreign government to organize foreign buyers in a way designed to increase its monopsony power (i.e., imposing offset requirements) vis-à-vis sellers seeking to penetrate that national market (or to exert that monopsony power directly, if the foreign government is the sole potential buyer) so as to increase some measure of national benefit. Would it also be desirable for the U.S. government to help organize U.S. sellers in ways that increase their bargaining power—and U.S. national benefit—to countervail foreign buyers organized in monopsony-enhancing ways?

Some such forms of monopoly-power-enhancing organization are already permitted. Voluntary private export cartels (domestic international sales corporations) are legal if certain requirements are met. Government-led efforts are unlikely to be successful, unless the government is funding much of the product development or technology export is regulated directly by the government. Both of these conditions, however, are met in the U.S. defense industry.

Of course, if there are foreign providers of the same systems that are not part of a U.S.-organized supplier's club or are not subject to U.S. government-imposed export restrictions, there is a good chance that a U.S.-only effort to impose limits on exports would fail, and the United States would simply give the business to foreign suppliers. On the other hand, in high-end military systems, there are a relatively small number of countries with advanced capabilities, most of whom are U.S. allies, and it is not impossible to envision some grand agreement on limits on technology transfer to support common security objectives, if some part of that agreement also addresses the ability of all parties to maintain needed defense-related industrial capabilities. This was the idea behind the "inner circle" concept mentioned above.

In commercial industries with limited numbers of international players, it is also possible to conceive of international arrangements that might benefit vendors facing buyer offset demands. Boeing and Airbus, for example, might in theory be sanctioned by their respective national antitrust authorities to set common maximal limits on co-production or technology transfer conditions attached to commercial sales deals to complement Euro–American national controls on defense aerospace technology exports.

Public Investments and Private Profits

One bright line separating defense offsets from commercial offsets is the role of governments in funding R&D. Although private firms are probably best equipped to secure the deals that capture the maximum return on *private* investments in new technology, the same may not hold true when it comes to securing the maximum national return on *public* investments in new technology.

For example, if \$15 billion is invested in developing a new engine technology, a firm may logically consider its direct return from licensing the technology and \$3 billion in lost profits on possible future sales won by its now more competitive foreign licensee and decide that \$4 billion in licensing fees is a good deal. If the company alone invested in the technology, that would be the end of the discussion. If the government funded the \$15 billion, however, and made the resulting know-how available to multiple U.S. companies, it might reasonably want a U.S. company to consider the possible costs of future competition to *other* U.S. firms as well. If this future loss from the new competition to *all* U.S. firms were, say, \$6 billion, the \$4 billion licensing deal would be considerably less attractive from a national perspective. This is a calculation that a U.S. company would not normally make in evaluating the deal from its own purely private perspective, but it might be the appropriate one in considering the transfer of know-how based on publicly funded R&D.

Although the solution to this problem is far from clear, it is equally certain that the terms on which publicly funded technologies are transferred or sold to foreign buyers by their private developers will continue to be the subject of vigorous public debate when those technologies draw from a common pool of knowledge extending beyond the boundaries of the seller. The government, as the custodian of the public interest in taxpayer-funded R&D, will continue to be called on to make decisions and defend them when export of technology is considered.

We should also accept that the bright line between military and commercial investments is often much fuzzier than is commonly acknowledged. Most commercial jet engines developed over the past 20 or 30 years, for example, were largely based on cores and other technologies originally developed for military platforms. Indeed, the jet engine business remains highly dependent on military-funded R&D for much of the continuing advance in its technology base, despite the fact that only one-fourth of the value of U.S. shipments today comes from military units.

The Trading Regime

Offsets might be regarded as a form of subsidy to exports (because other goods, services, and commitments with some economic value are being bundled into a sales transaction). There are restrictions on subsidies and pricing behavior in international trade that discipline the use of such subsidies, and governments therefore are interested in offsets as a trade issue in sectors where they may be used to promote exports by national companies. The defense sector (including much of aerospace) is unique in this regard in that the national defense "escape" clause written into the GATT exempts defense goods and services from some of the effects of these disciplines. The limits on "greenlighting" of R&D subsidies to product development in commercial sectors, for example, arguably do not apply to defense articles. Indeed, one might even argue that what might be labeled as "dumping" (sales of products at prices that do not cover the fully loaded—including R&D—cost of production) is routine practice in international sales of defense articles.

There is a real danger that offsets can be used as a backdoor around GATT trade disciplines. One example is the apocryphal (but generally accepted) account that has the Japanese government agreeing to use its good offices (and regulatory powers) to encourage Japanese airlines to purchase Boeing aircraft in exchange for Boeing's increased use of components made by Japanese aerospace suppliers. (Japan Air Systems reportedly defied the government pressure and purchased Airbus jetliners at a very good price, at the cost of a considerable negative campaign in the press and elsewhere against it.) In essence, the European supplier of Airbus became a "less favored nation" when selling into the Japanese market. In a similar view, a Boeing executive publicly acknowledged at a National Academy of Sciences forum in the early 1990s that Japan's MITI had cut a deal that involved assuring Boeing that Japanese aerospace companies would not team with Airbus to produce a new aircraft competitive with Boeing's line in exchange for work guarantees in producing components for Boeing aircraft. Both agreements could be interpreted as offset arrangements, although these agreements also have a certain "informal" quality that make them less transparent than examples normally raised in discussions of offsets.

The challenge to the spirit, if not the letter, of the GATT is obviously much more relevant to offsets in commercial sales because defense is given a wide berth by WTO disciplines. In the long run, the solution to this problem may well be an explicit definition of "informal" trade barriers covered by GATT principles that includes offsets and further agreements on transparency of procurement procedures in government-owned, regulated, or subsidized industries that ultimately defines government offset requirements as an explicit trade barrier.

But as noted above, the dividing line between defense and commercial products is not always sharp. Some might even argue that this line is growing ever wider and fuzzier as more and more industries historically dominated by military procurement (such as space) increasingly turn to commercial customers and financially strapped defense ministries turn to low-cost commercial alternatives for specialized defense components historically purchased from a captive military supplier base.

CONCLUSION

With defense downsizing in full swing around the globe, all major producers of high-technology armaments other than the United States face virtual economic crisis in their defense industries. Unless they are willing to give up maintenance of a national capability to produce advanced military systems as a national security objective (exceedingly unlikely), this means that they will be pushed to (1) close off their national markets to foreign-built systems and (2) dramatically increase exports. In the long run, this is likely to raise significant problems for the United States.

Offsets are a competitive tool used by both buyers and sellers in an increasingly cutthroat global market for defense equipment. The United States is alone in being able to sustain a viable, broad-based defense industry solely on the strength of its large domestic market; other countries with advanced arms industries must export substantial volumes of defense goods into a shrinking global market just to keep their industrial base economically sustainable.

Offsets create policy challenges in three ways. They offer a way for governments to insert themselves into private transactions to increase the bargaining power of domestic firms in international agreements or advance other government objectives. If successful, however, offset requirements may invite retaliation through a countervailing policy of organization of private or public cartels among suppliers. (This might even work in industries with small numbers of suppliers, such as defense and aerospace.) Second, their use as leverage to force technology transfer from private suppliers raises difficult policy questions about the extent to which private interest fully captures the public interest in technologies funded out of public coffers. Third, use of offset requirements is a backdoor challenge to the GATT vision of open global markets.

The obvious alternative in military aerospace that addresses both the big issue of unrestrained exports of advanced military capabilities and the smaller issue of offsets (one tool used to pry open greater technology transfer) is to work through some sort of system of industrial and technical cooperation with major U.S. allies (Europe and Japan). Some arrangement would be needed that would maintain access by U.S. defense producers to these important markets while permitting U.S. allies to maintain core defense systems capabilities and restrain the unchecked proliferation of advanced military exports. The United States—as the only nation that can maintain an economically affordable advanced defense sector without relying on exports—must play a leadership role in constructing such a system. The massive investment by the United States in military technology—which in effect underwrote the development of allied industrial capabilities in the first place—continues to provide us with enormous leverage that can be utilized for this purpose.

The Effects of Offsets, Outsourcing, and Foreign Competition on Output and Employment in the U.S. Aerospace Industry

Robert E. Scott
Economic Policy Institute

The debate over the employment effects of offsets is contentious because of the interplay of several closely related questions that can be quite difficult to disentangle. Overall employment has declined steeply in the U.S. aerospace industry since 1989 for a variety of reasons, including sharp reductions in public spending for defense goods and space exploration; rising imports of aircraft, engines, and components; and increasing productivity and structural changes in the defense and commercial aerospace industries, including numerous mergers and consolidations. Furthermore, industry representatives argue that, in the absence of offsets, foreign sales of commercial and defense aerospace equipment would decline or disappear, raising questions about the appropriate counterfactual exercise that should be used to analyze the effects of offsets and related issues on total industry employment.[1]

Barber and Scott (1995:2) examined a number of these issues and concluded that, in addition to the 500,000 jobs already lost in this industry since 1989, an additional 469,000 jobs in aerospace and related industries could be at risk by 2013 because of offset policies and increased foreign competition." In this report I update Barber and Scott (1995) and re-examine the evidence that increasing foreign competition will threaten a large number of jobs in this industry over the next one to two decades. I begin by reviewing trends in aerospace employment and then examine the principal causes of declining sectoral employment using a straightforward counterfactual exercise to assign changes in employment that

[1]See, for example, Johnson (1997:34) who noted that "if a foreign customer wants to deal in offsets, companies will have to listen and negotiate."

occurred between 1989 and 1996 to their proximate causes. In the next section I examine international competitive challenges and forecast the effects of offsets and other types of international competition on industry employment for the next two decades. The paper concludes with a discussion of policy alternatives for addressing the industry's problems.

EMPLOYMENT TRENDS

Overall Trends

Between 1989 and 1995 total employment in the aerospace industry declined by 545,000 workers, as shown in Table 1. In 1996 output and employment began to recover and they improved in 1997 (AIAA, 1997), and are forecast to increase for several more years (*Aviation Week and Space Technology*, 1997). The most important causes of the decline in employment were (1) declining defense budgets, (2) a worldwide recession in commercial aircraft demand, and (3) the effects of increased international competition. Between 1989 and 1995 overall aerospace employment declined by 40 percent. The decline in employment had very similar effects on production and nonproduction employment in the aerospace industry. The share of production workers in total industry employment fell only by 1.6 percentage points between 1989 and 1995, and the production worker share recovered strongly in 1997 (AIAA, 1997).

The global recession in aircraft demand caused employment to fall in all major aircraft-producing nations, as shown in Table 2. Total aerospace employment in the triad countries (the United States, Europe, and Japan) fell by more than 550,000 workers, according to the European Commission (EC).[2] However, the losses were not evenly spread. Employment in the United States and the United Kingdom, as reported by the EC, fell by about 42 percent between 1989 and 1995, but declined only 20 percent in the rest of Europe, 13 percent in Canada, and was unchanged in Japan. The United States absorbed about 74 percent of the job losses during this period, although only 62 percent of total triad aerospace employment was located in the United States in 1989.

As a result of these differential impacts, the U.S. share of aerospace employment in the triad countries declined by about 6 percentage points between 1989 and 1995, as shown in Figure 1. The other members of the triad all gained employment share, relative to the United States (with the exception of the United Kingdom, as shown in Table 2). Data reviewed below suggest that one reason that employment levels remained higher in Europe was the growing share of Airbus Industrie in commercial aircraft markets.

---

[2]The EC uses a narrower definition of the aerospace industry than does the Aerospace Industries Association of America, as shown in Table 2.

TABLE 1 Aerospace Industry Employment, 1982–1997 (in hundreds of thousands)

Year	Total Aerospace	Total Aircraft	Civil Aircraft	Military Aircraft	Missiles and Space	Other Related	Other Related Share of Total (%)
1982	1,027	516	231	285	243	268	26.1
1983	1,027	484	174	310	259	284	27.7
1984	1,097	517	184	333	286	294	26.8
1985	1,206	588	210	378	294	324	26.9
1986	1,272	639	238	401	309	324	25.5
1987	1,300	653	257	396	316	331	25.5
1988	1,311	666	280	386	313	332	25.3
1989	1,331	702	326	376	306	323	24.3
1990	1,270	687	341	346	281	302	23.8
1991	1,180	660	345	315	251	269	22.8
1992	1,040	597	322	275	217	226	21.7
1993	907	523	275	248	176	207	22.8
1994	827	469	285	184	165	193	23.3
1995	786	423	256	177	172	181	23.0
1996[a]	805	451	282	168	171	183	22.7
1997[b]	826	469	304	165	172	185	22.4

[a]Preliminary.
[b]Estimated.

SOURCE: Economic Policy Institute analysis of AIAA (1996, 1997).

TABLE 2 Aerospace Employment in Europe, Canada, Japan, and the United States, 1974–1995

Year	United Kingdom	Other European Union	Total European Union	United States[a]	Canada	Japan	Total
1974	210,100	199,541	409,641	666,000	28,400	29,814	1,133,855
1979	196,566	227,071	423,637	775,000	37,700	31,666	1,268,003
1980	229,821	241,874	471,695	830,000	46,800	32,991	1,381,486
1984	203,202	262,318	465,520	817,000	44,041	34,200	1,360,761
1985	206,677	274,971	481,648	898,000	48,794	34,300	1,462,742
1989	189,911	295,829	485,740	992,000	66,106	38,300	1,582,146
1990	186,337	297,635	483,972	946,000	65,679	39,100	1,534,751
1994	119,353	240,954	360,307	616,000	54,031	38,100	1,068,438
1995	110,549	237,512	348,061	580,000	57,329	38,300	1,023,690

[a]Figures for U.S. employment include only companies in SICs 372, 376, 366, 381, and 382 and exclude other aerospace-related companies and their employees.

SOURCE: European Commission (1994, 1997).

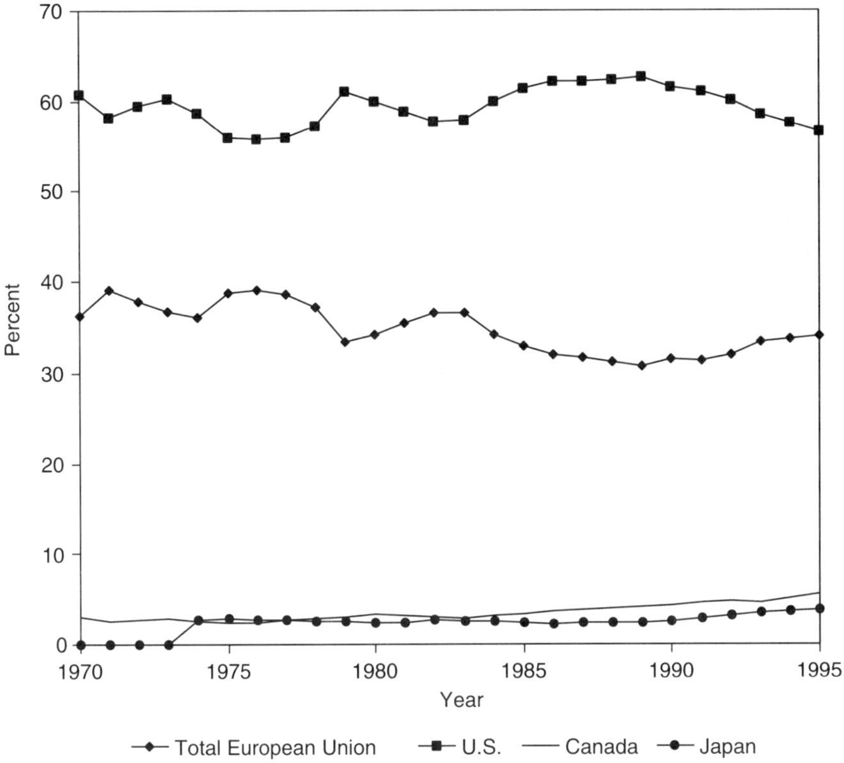

FIGURE 1 Shares of triad aerospace employment. Source: Economic Policy Institute analysis of data from European Commission (1997).

Causes of U.S. Aerospace Job Losses, 1989–1997

Industry revenues, measured in constant dollar terms, declined sharply in all the major sectors of demand in the early 1990s, as shown in Table 3. The sectoral changes were not evenly spread, as was the case with employment in Table 1. Total industry revenues fell 30 percent between 1990 and 1995 and increased 26 percent between 1995 and 1998. Civil aircraft sales fell by 32 percent in the first half of the 1990s, but are forecast to nearly double between 1995 and 1998. Missile sales declined much more rapidly than average and will remain flat through 1998. Space purchases (including research and development [R&D]) were essentially flat in the 1990s. Of the five major subsectors of the aerospace industry, shown in the last five columns of Table 3, civil aircraft sales were responsible for the greatest majority of the industry's $23 billion expected increase in real sales

TABLE 3 Aerospace Industry Revenues (constant 1987 dollars, in millions)

Year ($Millions)	Total Aerospace ($Millions)	Total Aircraft ($Millions)	Civil Aircraft ($Millions)	Military Aircraft[a] ($Millions)	Missiles[a] ($Millions)	Space[a] ($Millions)	Related Products and Services
1979	71,528	41,546	20,830	20,717	7,524	10,307	12,150
1987	110,008	59,188	15,465	43,723	10,219	22,266	18,335
1988	112,426	59,751	18,664	41,086	10,079	23,859	18,738
1989	113,604	58,011	20,644	37,367	12,839	23,821	18,934
1990	121,606	64,573	28,382	36,281	12,833	23,933	20,268
1991	121,508	66,246	32,673	33,573	9,572	25,438	20,251
1992	117,251	62,525	33,754	28,772	9,947	25,238	19,542
1993	101,636	54,314	27,323	26,991	6,973	23,409	16,940
1994	89,160	46,490	20,642	25,848	6,099	21,710	14,860
1995	85,473	43,654	19,005	24,649	5,857	21,717	14,246
1996	91,364	46,987	21,074	25,913	6,309	22,841	15,227
1997[b]	99,480	53,048	29,657	23,391	6,358	23,494	16,580
1998[c]	108,121	59,540	36,754	22,785	6,003	24,559	18,020
Change in Constant Dollar Revenues (%)							
1979–90	70	55	36	75	71	132	67
1990–95	–30	–32	–33	–32	–54	–9	–30
1995–98e	26	36	93	–8	2	13	26

[a]Includes funding for research, development, testing, and evaluation.
[b]Preliminary.
[c]Estimated.

SOURCE: Economic Policy Institute analysis of AIAA (1996, 1997).

TABLE 4 Relationship of U.S. Aerospace Revenues to Exports, Imports, and the Balance of Trade (billions of current dollars)

Year	Total Aerospace Revenues ($)	Total Aerospace Export Revenues ($)	Exports as Percent of Total Aerospace	Total Aerospace Import Revenues ($)	Imports as Percent of Total Aerospace	Aerospace Balance of Trade ($)
1979	45.4	11.7	25.9	1.6	3.6	10.1
1989	120.5	32.1	26.6	10.0	8.3	22.1
1990	134.4	39.1	29.1	11.8	8.8	27.3
1991	139.2	43.8	31.4	13.0	9.3	30.8
1992	138.6	45.0	32.5	13.7	9.9	31.4
1993	123.2	39.4	32.0	12.2	9.9	27.2
1994	110.6	37.4	33.8	12.4	11.2	25.0
1995	106.3	33.1	31.1	11.5	10.8	21.6
1996[a]	112.4	39.6	35.2	13.6	12.1	26.0

[a]Preliminary.

SOURCE: Economic Policy Institute analysis of AIAA (1996, 1997).

between 1995 and 1998. Increased sales of $3.8 billion in related products and services roughly offset losses in military sales in this period.

Imports have also been rising rapidly, as shown in Table 4. Between 1989 and 1997, imports increased by $8.1 billion in nominal terms. The import share of the domestic market increased by 5.7 percentage points. One way to assess the effects of trade on employment is to construct a counterfactual scenario in which trade remains constant as a share of domestic output. In this scenario, if demand were unchanged in 1997, then domestic sales and output could have been 5.7 percent higher than they actually were if the import share had remained unchanged at its 1989 level. This translates into about 49,500 jobs, out of the actual loss of 462,000 jobs between 1989 and 1997(see Table 1). Thus, increased imports directly account for about 11 percent of the decline in aerospace employment observed in this period.

Falling output also reduced employment, although a significant recovery had occurred by 1997, relative to the nadir reached in 1995. Total industry sales, in real terms, declined by 12.5 percent between 1989 and 1997 (Table 3). If output were unchanged in 1997, relative to 1989, then an additional 109,000 jobs would be been retained or created in this industry. Thus depressed demand for aerospace products explained about 24 percent of employment losses in this period. Declining demand in all sectors, especially in the military industries, was the most important cause of falling employment between 1989 and 1995 (as indicated in Table 3), but productivity and related factors played a more significant role after 1995.[3]

Productivity growth, broadly defined as output per worker, also eliminated significant amounts of aerospace employment in this period. The unexplained employment changes (those due to forces other than increased imports and demand changes) amounted to 292,000 jobs between 1989 and 1997, or two-thirds of all jobs lost in this period. However, this measure also includes the effects of changes in the capital/labor ratio, and interactions between productivity and the other factors discussed above (trade, demand, and capital intensity). In the absence of data on changes in capital inputs in this period it is impossible to estimate the effects on employment of total factor productivity growth, or pure technical change as it is usually measured by economists. Nonetheless, it appears that productivity growth does explain a substantial share of the job losses that did occur between 1989 and 1997.

In assessing the causes of declining employment, it is also important to note that trade also has indirect effects on employment that are not reflected in the

[3]Editing of the proceedings for this volume permitted updating of the data reported in this chapter. The analysis in this section has been revised to reflect changes in trade and output through 1997, as a result. These revisions altered the details of the analysis, but did not affect the basic conclusions. In particular, the trend rate of growth in the ratio of imported engines and parts to aircraft sales, a key indicator in the analysis in this chapter, accelerated sharply in 1997, as shown in Figure 4, below.

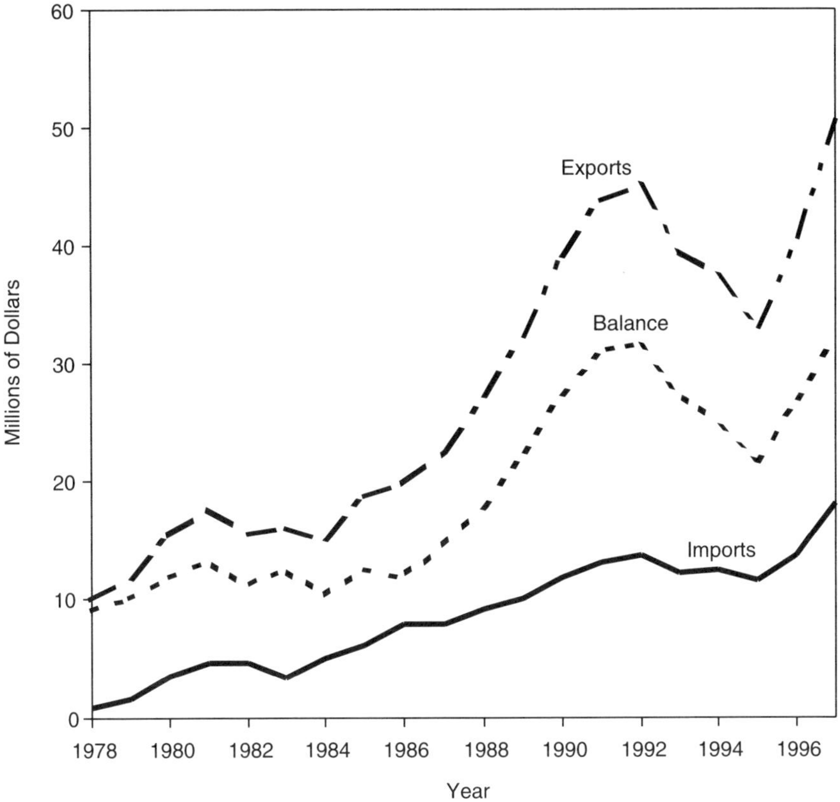

FIGURE 2 Aerospace exports, imports, and trade balance. Source: Economic Policy Institute analysis of AIAA (1996, 1997).

simple decomposition given above. For example, increased foreign competition may explain a significant portion of the observed productivity increase. In addition, the loss of foreign market shares also explains part of the decline in total sales. As shown below, Airbus, in particular, has sharply increased its share of foreign markets for commercial aircraft, which are some of the most important sources of demand for U.S. producers (Table 4).

The increasing international competition, and the global decline in aircraft demand, led to a sharp decline in the nominal value of U.S. exports between 1992 and 1995, as shown in Table 4 and Figure 2. Exports began to increase again in 1996, stimulated by the end of the commercial aircraft glut of the early 1990s and also by the introduction of Boeing's new 777 aircraft. The effects of the international business cycle were amplified by a technology cycle in which Airbus was first to market with its new generation of long-haul aircraft (the A330 and A340 models, first delivered in 1993) and was therefore able to capture market share from the U.S.-based producers until the 777 was launched in 1995 (European Commission, 1997:20).

Despite these difficulties, export sales fell less sharply in the early 1990s than other types of aerospace sales (Table 4). Therefore, because their share of domestic production increased, the constant share model suggests that export markets helped sustain domestic production in this period. Exports increased from 29.1 percent of total revenues in 1990 to 38.8 percent in 1997, an increase of 9.7 percentage points. Without this increase in export sales, U.S. aerospace employment would have been reduced by a similar amount in 1996. In other words, if the export share of U.S. aerospace sales had remained constant at its 1989 level, then there would have been approximately 84,000 fewer jobs created.

The question that must be addressed is whether a constant export share is the appropriate counterfactual exercise. The United States controlled a substantial share of the world commercial aircraft industry in the 1980s, and in fact that share has fallen in the 1990s, as discussed below. It is the growth in overall world demand for commercial aircraft, in combination with the shrinkage of demand for other domestic aerospace products, that has caused the increase in the export share of U.S. aerospace sales. For these reasons it is incorrect to assume that these export sales, and the increase in the share of exports in total sales, would not have taken place in the 1990s unless imports had also increased, as suggested by some observers. The two trends are driven by different forces. Changes in imports and exports should be analyzed independently.

FUTURE THREATS TO U.S. AEROSPACE EMPLOYMENT

Airbus and other potential new foreign competitors in the commercial aircraft industry are the most important direct threats to U.S. aerospace employment. This threat was realized and became increasingly important in the 1990s, as shown in Figure 3. The market share of Airbus and other European aircraft producers began to rise sharply, especially after 1992, in all significant markets. In the United States their share attained a peak of 30 percent in 1994 and has declined since. However, their share in the rest of the world (excluding the United States and Europe) increased by nearly 50 percent between 1994 and 1995, from 30 percent to nearly 45 percent. This reflects, in part, the technological cycle discussed above. However, it may also reflect (1) more aggressive sales promotions by European Union (EU) governments (for example, through more generous export financing), (2) more competitive pricing by Airbus, and (3) market-share gains through increased willingness to engage in offset transactions. In China, in particular, Airbus has increased its market share sharply and has also announced a number of joint and co-production arrangements with Chinese producers (see examples in Box 1).

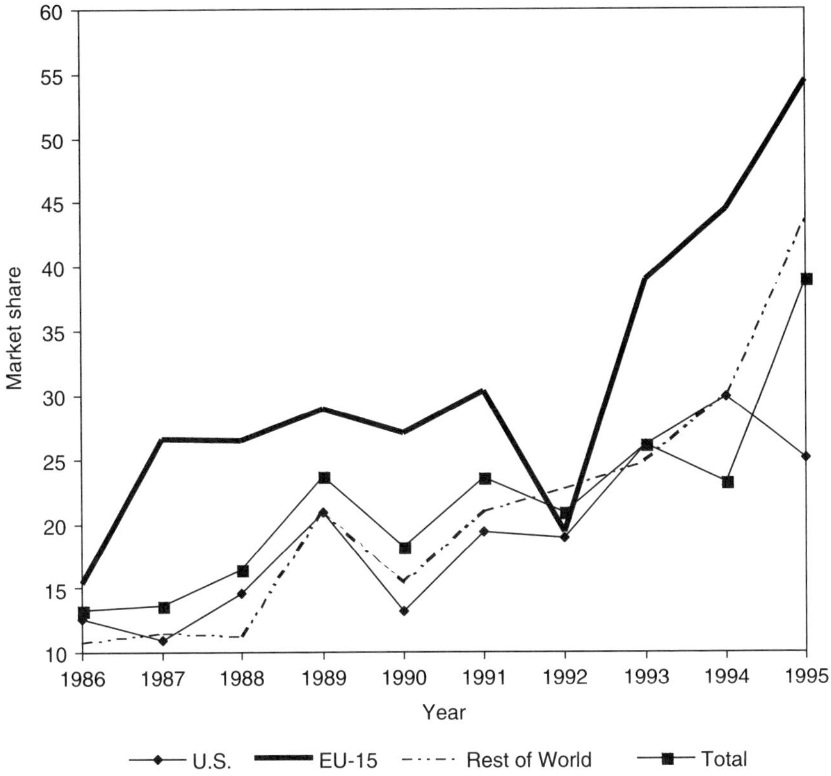

FIGURE 3 European Union shares of aircraft deliveries, by region of the world, 1986–1995. Source: Economic Policy Institute analysis of European Commission (1997).

Airbus and the other European aircraft manufacturers have also maintained a larger share in their home market throughout the period shown in Figure 3. They became the dominant suppliers in Europe in the 1990s. In 1995 European firms captured 55 percent of their home market (measured in value terms), while controlling approximately one-third of the total world market for commercial aircraft sales. In 1995, 22 percent of European commercial aircraft sales were in their home market (European Commission, 1997:32). Boeing is also heavily dependent on exports. It now "exports 70% of its commercial jets," a fraction that has been increasing in recent years (Waldmann, 1997:6). The relatively large market share of Airbus and other European aircraft producers in the EU suggests that European airlines (public, private, and mixed) may be skewing sales to European firms for nonmarket reasons. Although such preferences may violate the spirit or letter of the 1979 General Agreement on Tariffs and Trade (GATT) aircraft agreement (signed by the United States and the countries of the EU) or the 1992 civil

BOX 1
Offsets and Outsourcing in China

- Airbus donated a $50 million flight training simulator for use in training pilots in Beijing.
- Xian Aircraft Co. builds vertical fins and horizontal stabilizers for the Boeing 737, trailing edge ribs for the 747, and forward access doors for the 737.
- Xian also builds access doors for the Airbus A300, A310, A330, and A340 and carbon fiber fin ribs for the A320.
- Chengdu Aircraft Co. makes vertical fins, horizontal stabilizers, and tail sections for the Boeing 757 and nose sections for the MD-80 and MD-90 transports.
- Shenyang Aircraft Co. makes cargo doors for the Boeing 757 and wing ribs and emergency exit doors for the Airbus A320 and machine parts for the A300 and A310.
- Shanghai Aircraft Co. makes components for the Boeing 737 and jointly assembles the McDonnell Douglas MD-80 and MD-90 transports.

Source: *Seattle Times* (1996).

aircraft accord, they have not yet been the subject of formal complaints or action by U.S. producers or government agencies such as countervailing duties or Section 301 actions.

The trends illustrated in Figure 3 are reflected in Table 5, which reports U.S. aerospace exports, by region, between 1991 and 1995. Exports to the eight largest European countries fell 40 percent, the largest decline of any region. This reflects the particularly strong performance of European firms in their own market in the 1990s. U.S. exports to Japan and Asia were essentially flat, although they did decline somewhat from 1992. Exports to the rest of the world declined by 25 percent in this period, also reflecting the effects of increased competitiveness of Airbus, vis-à-vis Boeing and McDonnell Douglas, although it is not clear whether this reflects market forces or unfair competition.

In the context of the offsets debate, in which China is often cited as one of the most aggressive and egregious violators of international norms, and also one of the most important and rapidly growing markets for aircraft, it is significant to note that total U.S. aerospace exports to China were quite small, representing between 3.5 and 7.5 percent of total U.S. exports in this period. The small volume of U.S. exports to China suggests that the benefits of offsets, in terms of increased sales, may be quite limited. However, the costs in terms of lost jobs and diffusion of critical technologies could be quite significant. Ultimately, foreign aerospace

TABLE 5 U.S. Aerospace Exports, by Region (millions of dollars)

Year	EU-8	Japan	China	Rest of World	Total Exports
1991	17,647	3,910	1,244	12,747	35,548
1992	14,801	4,505	2,247	15,353	36,906
1993	11,802	3,581	2,384	14,056	31,823
1994	11,821	4,099	2,047	12,083	30,050
1995	10,620	3,587	1,250	9,622	25,079

SOURCE: Economic Policy Institute analysis of AIAA (1996).

firms may play a key role in helping China, or a coalition of Asian aerospace firms, become a new entrant in the aircraft industry as a full-fledged designer and integrator of commercial jets, with potentially devastating consequences for both Airbus and Boeing.

Offsets as Threat to Future U.S. Aerospace Employment

Imports represent a growing share of domestic output, as noted above in the discussion of Table 4 on U.S. aerospace sales and trade. Imports have two distinct effects on output. First, an increase in the sales of finished aircraft of foreign origin reduces the level of domestic sales. In this sense, imports directly represent an opportunity cost in terms of lost sales and employment possibilities.

The second channel through which imports can affect domestic employment is when the use of foreign parts and components increases as a share of value added. In effect, the "foreign content" of domestic sales is thereby increased, and a given volume of final sales will not support the same number of domestic jobs. This is equivalent to a reduction in the amount of labor that is required to produce a given amount of output.[4] Offset agreements, both voluntary and mandatory,

[4]If input-output relationships are used to model the effects of trade and output on domestic employment, then an increase in the share of imported components in final sales should be reflected in a reduction in the direct and indirect labor requirements per dollar of value added. Note, however, that tables 2 through 4 report industry sales. This reporting procedure necessarily results in double counting of some elements of value added. For example, aircraft parts sold to final integrators are also included in the value of the aircraft finally delivered to domestic or foreign customers. It would be highly preferable if industry were to report domestic value added, rather than total sales.

[5]The exception to this rule would occur at later stages in the product life cycle when a foreign competitor begins to assemble finished aircraft. In this case, the aircraft assembled abroad would reduce domestic employment by either (1) reducing foreign (export) sales of the domestic firm and/or (2) increasing imports of finished aircraft. McDonnell Douglas has already reached both stages in its

will generally have the largest effect on parts and components imports.[5] It is therefore important to distinguish imports of components and complete aircraft when analyzing the effects of trade on employment.

In the 1990s, imported parts and components (including engines and engine parts) were, on average, 71 percent of total aerospace imports.[6] Figure 4 shows that U.S. engines and parts imports represent a growing share of total aircraft sales.[7] This share rises from about 8 percent in 1981 to over 19 percent in 1997. Between 1984 and 1997, this share increased by about 0.7 percentage points per year. As a first approximation, the increase in parts imports reflects the direct effects of offsets and all other forms of co-production and international outsourcing agreements. This measure is separate from the effects of increasing international competition in final aircraft sales, as reflected in increased imports of complete aircraft and reduced exports. This distinction will be important in predicting future jobs at risk.

Before turning to the analysis of future employment threats, we need to examine trends in some of the major components of aircraft employment. Figure 5 reports total employment in airframes (final assembly and integration), engines and parts, and other parts and equipment. Several trends stand out in these data. First, the bulk of employment in this industry is still in the airframe sector, composed primarily of the major defense and commercial aircraft integrators. Second, there appears to be a secular declining trend in employment in airframes, engines, and parts. As a location for production, the United States is internationally the least competitive in the engine sector. The United States has had roughly balanced trade (exports equaling imports) in this sector in recent years and has periodically experienced overall deficits in engine trade. This is remarkable, in part, because two of the three major engine producers are based in the United States (Almeida, 1997).

---

co-production arrangement involving Chinese assembly and co-production of its MD-82 and MD-90 aircraft. At least one of these aircraft has also been exported back to the United States.

When final aircraft assembly is transferred abroad, parts exports will increase even while total industry employment and value added would decline, relative to what they would have been if those aircraft had been assembled in the United States. In this case, it is inappropriate to treat parts and component exports as net-job creators. Indentification of appropriate counterfactual analysis in such cases will require detailed knowledge of offset sales contracts and whether or not such sales would have taken place in the absence of voluntary or mandatory offset agreements.

[6]There was no significant trend in the level of parts imports as a share of total aerospace imports in this period. This is somewhat surprising because of the decline in military production (which involves high levels of mandatory exports) and the relative increase in commercial aircraft markets noted above. Thus, the level trend in this share suggests that commercial sales may have involved increasing offsets in this period.

[7]Note that Figure 4 uses a different basis (total aircraft shipments) from the analogous exhibit in our prior research (Barber and Scott, 1995: Figure 8, p. 27). This change was made to improve the accuracy of our forecasts, as explained in the next section.

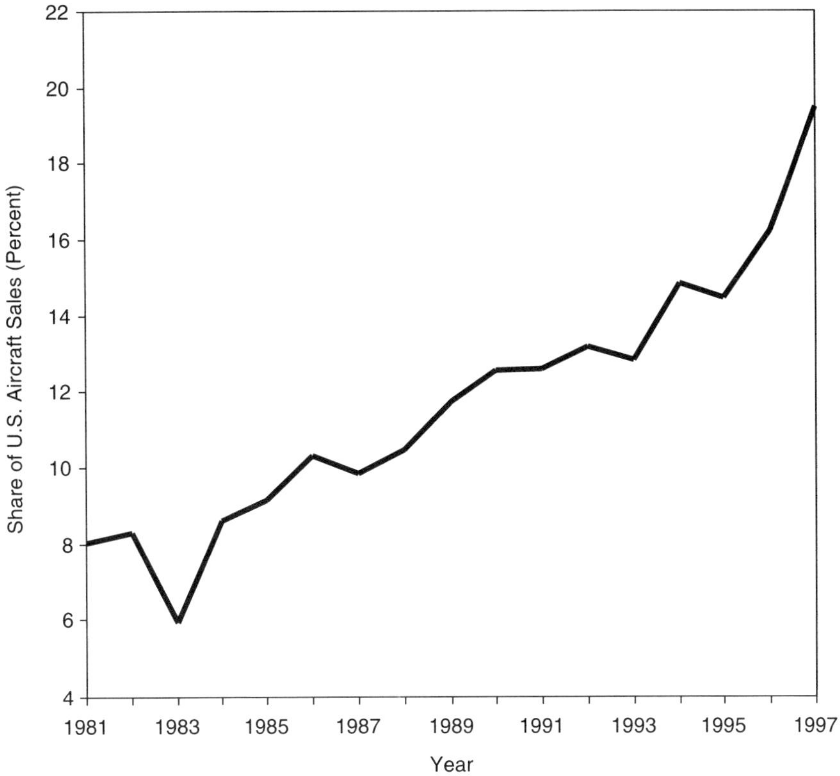

FIGURE 4 U.S. engines and parts imports as a growing share of total aircraft sales. Source: Economic Policy Institute analysis of AIAA (1996, 1997).

Third, there is some evidence of a stable or increasing trend in employment in the U.S. aircraft parts industry. This is illustrated in Figure 6, which compares employment in parts with employment in the much larger airframe sector. This comparison also reveals that employment in parts is somewhat more volatile than in the airframe sector. There are several possible explanations for these trends that are consistent with the evidence on rising parts imports shown in Figure 4. First, as the technological complexity of aircraft has increased it is likely that the number and value of components (electronics, for example) has also increased, as a share of value added. Second, modularization of component production has eliminated final assembly labor in a wide range of industries, from motor vehicles to electronics, so it is not surprising to observe evidence of that trend here. Finally, products counted as parts may substitute for production that used to take place within the plants of airframe manufacturers. Several examples of this are given in Box 1 that involve Chinese offsets and outsourcing, including doors,

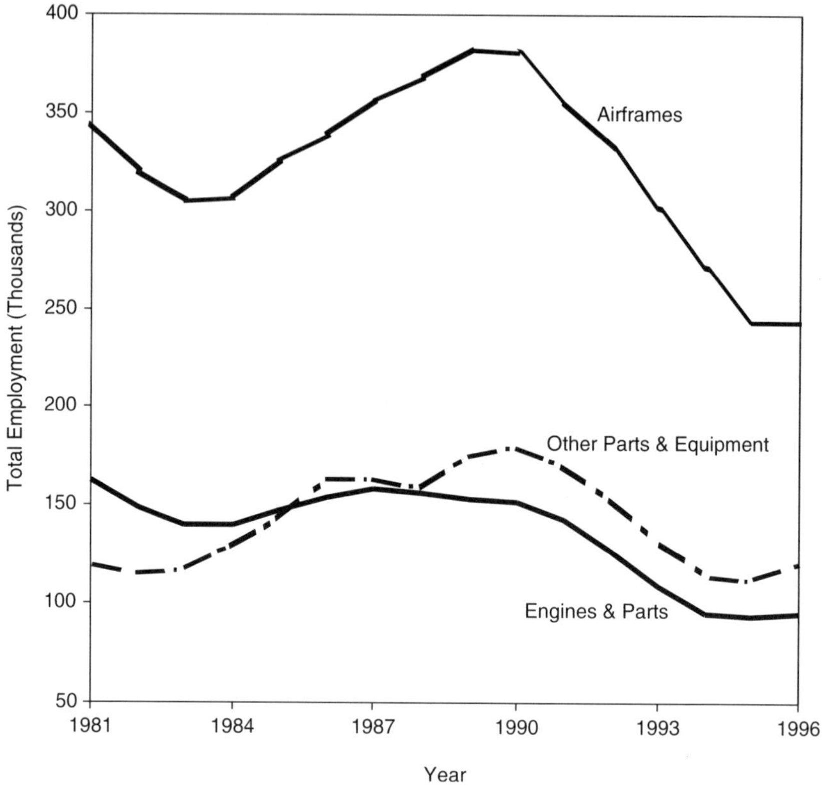

FIGURE 5 Long-run decline in employment in aircraft production. Source: Economic Policy Institute analysis of Bureau of Labor Statistics (1997) and AIAA (1996, 1997).

fins, stabilizers, and body parts for various types of commercial aircraft. As a consequence of each of these trends, employment in airframe manufacturing appears to be declining, relative to both domestic and foreign parts production. Thus, domestic workers in the airframe industry are correct to blame both outsourcing and imports for job loss in their industry.

The higher level of variability in parts employment suggests that employment in parts production is more heavily affected by the business cycle than is employment in the airframe sector. The secondary or sub-tier supplier base therefore appears to absorb a disproportionate share of the job losses that occur during downturns, once we control for the long-run shift of employment out of engines and airframes, as discussed above.

FIGURE 6 Employment in aircraft parts versus airframes. Source: Economic Policy Institute analysis of Bureau of Labor Statistics (1997) and AIAA (1996, 1997).

Jobs at Risk in the Future

In this section, employment effects of growing foreign competition, and also foreign outsourcing and other practices that encourage U.S. firms to shift component and parts production abroad, are briefly re-examined. As noted by several other contributors to this volume and other participants in the National Research Council conference (Wessner and Wolff, 1997), offsets are an important threat to domestic production in both military and commercial aircraft production. These problems are particularly important in the case of trade with large, nonmarket economies because of the role of and bargaining power exercised by government officials as the ultimate arbiters of all major import purchasing decisions. Some examples of current offset and co-production practices are illustrated in Box 1.

Boeing officials have acknowledged that their company does enter into voluntary offsets "as a means of gaining market access" (Waldmann, 1997:6). Other analysts concluded that, although the issue of offsets is "not even on the radar screen" at the present time, offsets will increase and become a "major factor over the next decade or two" (Bozdogan 1997:27). These views are confirmed by closely examining the trends discussion above and their implications for future employment in the U.S. aerospace industry.

International competition threatens U.S. production in at least two distinct ways. First, the market share of U.S. producers of commercial aircraft could decline in the future as a result of market-share gains by Airbus Industrie and/or other potential new entrants to this market.[8] This will impact U.S. employment primarily by reducing U.S. exports because the majority of commercial aircraft produced in the United States are now sold in export markets, as noted above.

The second threat to domestic employment is through increases in the foreign content of domestically produced aircraft as a result of increases in offsets and other forms of international component outsourcing. Ultimately, such outsourcing could help create new competitors to U.S. producers and hence exacerbate the first type of international competition. However, such secondary effects are ignored here and only the effects of increases in the foreign content of domestically produced aircraft on U.S. aerospace employment are considered.

The foundations for my estimate of the number of jobs at risk in the future because offsets increase foreign competition are (1) the 1994 DRI/McGraw-Hill study of the U.S. manufacturer market share, as reported in Barber and Scott (1995:43–45) and (2) Boeing's estimates of the future world market for commercial aircraft. These estimates are used to forecast the future effects of increased foreign competition on total (constant dollar) sales of U.S. aerospace products.

The potential job losses that could result from offsets and other types of foreign outsourcing are estimated by extrapolating the current trend in the rate of growth of aircraft engines and parts (see Figure 4) as a share of the value of total domestic aircraft sales (military and commercial). I assume that real output remains constant at the level that prevailed in 1994 and that the trend in the rate of growth in the imported parts share continues through 2013.[9]

A recently released (Department of Labor, 1996) input–output-based employment requirements table is used to estimate the direct and indirect employment effects of trade for both types of employment threats. These procedures are similar to those used in Barber and Scott. However, they differ in two respects. First, use of the new input–output table provides industry-specific employment

[8]This analysis ignores competitive threats in the military and space sectors. Increased competition in any of these other sectors could further reduce domestic employment opportunities in the aerospace industry.

[9]The year 1994 is taken as a base for comparison with the estimates developed in Barber and Scott (1995:40–47).

TABLE 6 Potential Job Losses Attributable to Increased Foreign Content of Aircraft Made in the United States and Total Jobs at Risk Due to Foreign Competition, 1994–2013

	Revenue ($ billions)		Total Job Losses		
Period	Peak Annual Lost Revenue	Cumulative Lost Revenue	Total Jobs	Direct Jobs	Indirect Jobs
1994–1998	1.6	4.0	17,863	10,219	7,644
1999–2003	3.5	17.8	39,498	22,596	16,902
2004–2008	5.4	41.1	60,386	34,546	25,840
2009–2013	7.2	73.5	80,553	46,083	34,470
Projected Losses Due to Declining U.S. Market Share					
	Average Annual Lost Revenue				
1994–1998	2.1	10.7	23,933	13,692	10,241
1999–2003	4.2	31.8	47,194	26,999	20,195
2004–2008	7.3	68.4	81,863	46,833	35,031
2009–2013	12.0	128.6	134,650	77,031	57,619
Total Jobs at Risk Due to Increased Foreign Competition					
1994–1998	3.7	14.7	41,796	23,911	17,885
1999–2003	7.8	49.6	86,692	49,595	37,097
2004–2008	12.7	109.5	142,249	81,378	60,871
2009–2013	19.2	202.1	215,202	123,114	92,089

SOURCE: Economic Policy Institute analysis of DRI/McGraw-Hill (1994: 9–11) and Boeing Company (1994:Appendix C).

multipliers that are much more accurate (and significantly smaller) than those used in the prior study. Second, the share of imported parts and components in aircraft sales is used to estimate the effects of outsourcing. This is a much more precise and appropriate base than the one used in our previous research (trends in total imports as a share of total aerospace sales).

The resulting projections of output and job losses are shown in Table 6. The first panel in the table reports the effects of increased foreign outsourcing, assuming that the trend of increasing foreign content of the past decade continues into the future. The first column of results shows the annual loss in output, relative to the base year of 1994 in constant dollars.[10] Employment losses are calculated, using the multipliers described above, at the end point of each period. Cumulative losses in output are shown for informational purposes only and are not directly used to calculate employment effects. The employment losses shown are the permanent losses of job opportunities, based on the given annual reductions in domestic output. More than half of all jobs resulting from a given amount of expenditure on aircraft are located in the aircraft and parts industries (direct jobs). Within ten years (by 2003) of the base year, 22,596 aerospace jobs will have been eliminated by outsourcing, plus an additional 16,902 indirect jobs in industries that provide inputs to the aerospace industry, for a total loss of 39,498 jobs. Within two decades, offsets and other forms of foreign outsourcing could eliminate 46,083 aerospace jobs and 34,470 jobs in other industries for a total loss of 60,553 jobs. The direct jobs lost in 2013 would equal 9.6 percent of total aircraft employment in 1994.[11]

This estimate of job loss may be too small for several reasons. First, Bozdogan (1997) and others appear to suggest that the growth of offset activity could accelerate in the future. Second, these estimates do not include any "indirect offsets" in other aerospace or nonaerospace sectors. However, recent reports from the U.S. Department of Commerce (Department of Commerce, BXA, 1996) suggest that indirect offsets are increasing as a share of total offsets, at least in the military sector. Finally, the base of production in 1994 is quite low because the recovery in aircraft production had not yet occurred (see Table 3). If a later base year were assumed, or if we assume that the trend rate of growth in output is positive, then job losses as a result of outsourcing will increase. On the other hand, some suggest that the level of outsourcing is constant or leveling off (Waldmann, 1997). If so, then the outsourcing estimates in Table 6 may be too high.

The threat of reductions in the U.S. share of the world commercial aircraft market could have an even larger effect on employment than outsourcing in the next two decades, as shown in the second panel of Table 6. In this case the U.S.

---

[10]The U.S. market share trend line was derived from historic data (1970—1992), combined with DRI/McGraw-Hill market share projections (1994—2000) and then extrapolated to 2013. Estimated lost commercial jet aircraft sales are derived from the difference between the 1989—1993 average U.S. revenue market share (73.5 percent) and the projected market share described above, using Boeing forecasts for constant 1994 dollar global revenues through 2013. Estimated job losses are derived from Bureau of Labor Statistics (1996) input-output data on the jobs supported by final demand for aircraft, expressed in 1987 dollars for the year 1993, which reports that 13,748 total jobs (7,865 direct and 5,883 indirect) were supported in 1993 by each $1 million (1987 dollars) in final demand for aircraft. Comparable estimates, expressed in constant 1994 dollars (adjusted for changes in aerospace prices between 1987 and 1994) are 11,184 total jobs (6,398 direct and 4,786 indirect) per $1 million of final demand (1994 dollars). These figures were then multiplied by projected revenue losses, from the 1994 base, to estimated lost jobs, as described in the text.

[11]The base of comparison in this example includes only those workers employed in aircraft and parts (Standard Industrial Classification 372).

share of the world market for commercial aircraft, which has been declining since at least 1970, could fall from approximately 80 percent in 1990 to 50 percent in 2003 and 35 percent in 2013, using the assumptions made by Barber and Scott (1995: Figure 11, p. 44). The new multipliers described above are used to estimate the employment effects of the output losses implied by these forecasts. By 2003, 26,999 direct jobs and 20,195 indirect jobs could be lost, for a total of 47,194 jobs lost in aerospace and related industries. Within two decades, 77,031 direct jobs and a total of 134,650 direct and indirect jobs could be lost because of declining U.S. shares of the world market.

The effects of outsourcing and potential losses of international market share are combined in the last panel of Table 6. By 2003, 49,595 direct and a total of 86,692 direct and indirect jobs could be lost in aerospace and related industries. By 2013, these totals increase to 123,114 direct and 215,202 total jobs lost. The direct jobs lost in 2013 would represent 25.6 percent of the total jobs in aircraft production in 1995. Although foreign competition will not eliminate the U.S. aerospace industry in the next two decades, it could greatly undermine employment in this critical high-skill, high-wage industry that, for at least four decades, has been one of America's leading export sectors.

These estimates of jobs at risk hold domestic output and labor requirements constant at levels prevailing in the mid-1990s. Other factors could increase or decrease actual employment in the aerospace industry. Given these assumptions, productivity growth would also reduce employment in the industry. On the other hand, if aircraft and parts exports continue to grow, they would put upward pressure on employment.[12]

This paper does not forecast the impact of changes in overall demand, or of productivity growth. However, it is likely that overall U.S. aerospace employment will follow a declining trend in the future. Both productivity growth and increased foreign competition will put substantial downward pressure on employment. Despite the strong upsurge in aerospace exports and revenues between 1995 and 1997, based on data in Table 3, aircraft sales and industry revenues remained 18 and 22% below their previous cyclical peaks (in constant 1987 dollars) in the early 1990s. The sharp increase in EU market shares between 1992 and 1995 (Figure 3) and the continued improvement in Airbus shares of aircraft sales in 1997, suggest that the U.S. share of the commercial aircraft market may be declining more rapidly than was assumed in Table 6. The surplus in aircraft parts, reflects, in part, aftermarket sales of the existing base of aircraft, which is dominated by U.S.-made planes. Rapid growth in the global stock of Airbus aircraft, and the retirement of older U.S.-made models, will begin to erode this surplus within the next five to ten years.

---

[12]The need to consider productivity and output growth was noted by David Mowery, who also pointed out that the U.S. trade surplus in aircraft parts and equipment has grown for the past 8 years.

For these reasons, U.S. aerospace employment is likely to decline in the future. Offsets, outsourcing, increased foreign competition and productivity growth will all contribute to these declines. While it is not possible to say which of these factors will be most important, they are all related to the decline of the U.S. as a competitive location for aerospace production, and the falling market shares of U.S.-based aircraft integrators. Expected future job losses in the aerospace industry can only be avoided or reduced through sharp improvements in the competitiveness of the U.S., and of aerospace firms based in this country.

CONCLUSIONS AND POLICY ALTERNATIVES

Other countries are actively targeting the commercial aircraft industry. The Chinese and other Asian governments are using trade and industrial policies to capture production and technologies from the United States. Many of these systems were developed with public support. The U.S. aerospace industry stands at the edge of a precipice. If the challenges it faces are not addressed, at least 215,000 additional jobs in aerospace and related industries will be eliminated over the next two decades.

In Barber and Scott (1995) we develop a broad range of policy recommendations that are needed to restore and maintain the international competitiveness of the U.S. aerospace industry. These include:

- creation of an aerospace executive and an interagency task force within the National Economic Council;
- promotion of aerospace production and employment through reform of government regulatory processes, including those that encourage firms to engage in offsets, and new programs to stimulate domestic demand for aerospace research and products; and
- negotiating fair international trade agreements, including new initiatives to bring China into compliance with the GATT codes and their own Memos of Understanding and other agreements with the United States.

These policies, taken as a whole, could constitute the initiation of a coordinated industrial policy for the aerospace sector. The United States is unique among developed and many of the larger developing nations in its refusal to develop and implement conscious and coherent industrial policies for the aerospace industry. The United States clearly has industrial policies for this sector, but they are uncoordinated, incoherent, and frequently internally contradictory.

Discussion and debate over the past two years, as reflected in the National Research Council conference (Wessner and Wolff, 1997) and in other papers in this volume, clearly suggest that a new national system for monitoring commercial offsets is urgently needed. This system should encompass both mandatory and voluntary offsets.

Additional measures are also needed to address the causes of offset and outsourcing problems. Defense demand has apparently stabilized, and the industry has begun to recover from a global aircraft glut, but the recovery has had only a small positive effect on industry employment. Foreign competition is now the most important future threat to domestic aircraft demand, as shown above. Rising imports of engines, parts, and components, in particular, are the result of offset deals that Boeing and McDonnell Douglas have made with foreign governments to trade aircraft sales for production and technology transfer. Offsets are the result of a prisoners' dilemma problem. In China, for example, there is effectively only one major buyer in this rapidly growing market (the government), and all three major assemblers are competing for sales by offering offsets. These companies are giving away more technology and production than would be warranted if the market were competitively structured.

Controlling Offsets

U.S. trade problems could be greatly improved through further agreements with the EU. The United States should open negotiations to revise the 1992 Aircraft Agreement with the EU, which technically expired in 1994 when it was excluded from the Uruguay Round Trade Agreements. In addition to reopening the subsidies provisions (which U.S. manufacturers feel are far too weak), there is an opportunity to fix the prisoners' dilemma problem from the supplier's side by prohibiting the export of jobs and technologies in exchange for sales. The EU may also be motivated to participate in such talks because of unresolved concerns about the merger of Boeing and McDonnell Douglas. U.S. firms have opposed a unilateral limit on offsets because they believe it would result in lost sales to Airbus. A new U.S.–EU agreement could benefit both Boeing *and* Airbus by restoring a competitive balance to this industry.

One precedent for government regulation of unfair sales practices is the Foreign Corrupt Practices Act (FCPA) of 1977, which was adopted after the discovery of a $12 million bribe paid in a sale of Lockheed L-1011 aircraft. Despite harsh penalties (both fines and imprisonment), the temptation to cheat may still be irresistible. It was recently reported that General Dynamics paid a $100 million bribe for a Korean military aircraft sale. One problem with the FCPA is that it was imposed unilaterally, and only on U.S. firms. The FCPA approach can serve as a model, but it must be modified to solve the aerospace trade problem.

The solution to the prisoners' dilemma is to impose a new marketing limit (no offsets in exchange for sales) *multilaterally*. The United States and the EU should adopt essentially identical measures proscribing such behavior by firms based in each region. This agreement would reduce the trade and jobs-distorting use of offsets worldwide. This measure should be acceptable to both U.S. and European aircraft manufacturers, because all would experience higher levels of future sales and production than they would without the agreement.

Defining the National Interest

Clearly, the United States has a national interest in aircraft sales and the movement of production and technologies abroad, which is different from that of the aircraft companies. Boeing and McDonnell Douglas may be content to become "virtual companies," designing planes but outsourcing most or all of their production. Stockholders' interests may be protected by collecting royalties on technologies developed with public support in the United States. However, U.S. workers and suppliers will clearly suffer.

Decisions about key aerospace technologies should be designed to protect the national interest in jobs and exports as well as corporate profits. All three can benefit from prompt, effective government action if the United States does not wait too long. If it does, aircraft production could go the way of color TVs and DRAMs (dynamic random access memories), and with it, the U.S. trade balance and future standard of living. If the United States acts quickly, it can retain and rebuild a strong aerospace industry for the twenty-first century and the high-wage, high-skills jobs that go with it.

REFERENCES

AIAA (Aerospace Industries Association of America, Inc.) 1996 and earlier editions. *Aerospace Facts and Figures*. Washington, D.C.: AIAA.

AIAA (Aerospace Industries Association of America). 1997. *1996 Year-end Review and Forecast UPDATE*. Washington, D.C.: AIAA.

Almeida, B. 1997. Are Good Jobs Flying Away? U.S. Aircraft Engine Manufacturing and Sustainable Prosperity. Unpublished manuscript. Center for Industrial Competitiveness, Lowell, Mass.

Aviation Week and Space Technology. 1997. "Robust 1996 sales for Boeing, Airbus," January 13, 146(2):371.

Barber, R., and R.E. Scott 1995. *Jobs on the Wing: Trading Away the Future of the U.S. Aerospace Industry*. Washington, D.C.: Economic Policy Institute.

Boeing Company. 1994. *Commercial Market Outlook, 1994*. Seattle, Wash.: Boeing Company.

Bozdogan, K. 1997. Is anyone there? Monitoring U.S. strategic interests. In C.W. Wessner and A.M. Wolff, eds, *Policy Issues in Aerospace Offsets: Report of a Workshop*. Washington, D.C.: National Academy Press.

Bureau of Labor Statistics. 1996. *Employment Outlook: 1994–2005 Macroeconomic Data, Demand Time Series and Input Output Tables*. Office of Employment Projections. Washington, D.C.: U.S. Department of Labor.

Department of Commerce, Bureau of Export Administration. 1996. Offsets in Defense Trade. May, U.S. Government Printing Office.

DRI/McGraw-Hill and the Teal Group. 1994. *High-Skill, High-Wage Production Jobs in the U.S. Aviation Industry*. Lexington, MA: DRI/McGraw-Hill.

European Commission. 1994 and 1997. *The European Aerospace Industry: Trading Position and Figures*. Brussels, Belgium: European Commission, Directorate-General III.

Johnson, J. 1997. Maintaining high value-added exports amidst structural change. In C.W. Wessner and A.M. Wolff, eds., *Policy Issues in Aerospace Offsets: Report of a Workshop*. Washington, D.C.: National Academy Press.

Seattle Times. 1996. How Boeing Woos Beijing. Stanley Holmes, May 26, p. A1.

Waldmann, R. 1997. Strategies for success in the commercial aircraft market. In C.W. Wessner and A.M. Wolff, eds., *Policy Issues in Aerospace Offsets: Report of a Workshop*. Washington, D.C.: National Academy Press.

Wessner, C.W., and A.M. Wolff, eds. 1997. *Policy Issues in Aerospace Offsets: Report of a Workshop*. Washington, D.C.: National Academy Press.

Offsets in the International Marketplace: An Aerospace Industry View

Joel Johnson
Aerospace Industries Association

The issue of offsets, a practice in which a purchasing entity, usually a government, demands that a seller not only provides a service or product, but in addition helps the purchaser to obtain additional technology, business, or investment, has sporadically captured political attention in Washington for nearly two decades. Critics of the practice express concern that offsets send work off shore and can create future competitors. Exporting companies generally respond that offsets are a necessary part of competing in the world market and have relatively little adverse impact on the economy, especially compared to the benefits that accrue from the sales themselves. A series of government studies and reporting requirements seem not to have changed many minds nor resulted in any serious new government initiatives.

Offsets, both formal and implicit, are particularly prevalent in the aerospace industry. Aerospace tends to be seen by industrial and industrializing countries as a key industry for several reasons. It is intimately related to a nation's security, with aircraft generally accounting for the largest share of defense equipment expenditures for any modern military. The industry is seen as a technology driver, including in manufacturing techniques, that also pulls along other high-technology sectors such as electronics, advanced materials, and sensors. Aerospace is viewed as a prestige industry—a hallmark of a modern economy. Every country with an industrial base seems to want to be a player in aerospace.

Military aerospace products are invariably purchased by governments. Even commercial aerospace products, particularly in developing countries, are often acquired by government agencies or state-owned enterprises, such as air traffic control organizations, airports, and national airlines. These acquisitions are a highly visible use of taxpayers funds to purchase a product from off shore. Politicians naturally dislike such transactions. To make them more palatable, it is helpful if such purchases can be linked to job creation and increased capability in the aerospace and defense sector. In short, when the taxpayers' funds are spent, they, or at least their political leaders, want a "piece of the action." Offsets are the inevitable result.

Thus it is not surprising that a discussion of offsets quickly turns to a discussion of the aerospace industry. A few notes about the U.S. aerospace industry might be helpful to put the offset issue in context. Aerospace has for many years been among the most dynamic and expansive of U.S. industries. In 1997 domestic and international sales by U.S. aerospace companies were estimated at $130 billion, or about 3 percent of all U.S. industrial manufacturing activity. New orders for the year totaled about $115 billion, and the backlog of orders at year end amounted to $218 billion. The industry currently employs approximately 870,000 Americans.

The industry's export performance has been most remarkable, particularly when compared with that of other U.S. industries. In 1997 exports totaled $50 billion, whereas imports of aerospace products reached about $16 billion. This means that the U.S. trade surplus in aerospace products was roughly $34 billion, a continuation of a long-term trend of positive trade balances.

As positive as these numbers are, it should be noted that, except for trade figures, all are lower than the last industry peak in the late 1980s and early 1990s. In 1989 the industry employed 1.35 million people, and its sales reached $139 billion in 1991. Two major trends hit sales and employment beginning in the late 1980s. First, the markets for aerospace products declined, as financially troubled airlines reduced new aircraft purchases and, with the end of the Cold War, the U.S. defense budget, particularly the procurement portion, plummeted. Second, the aerospace industry began a major restructuring, partly in response to lessened demand, but also to reduce costs by taking advantage of increased technological capabilities. Such technologies include use of computers to allow primes and subcontractors to design products without the use of draftsmen and extensive use of mockups. Improved machine tools and testing devices have reduced the work hours needed for production, testing, and rework. The bottom line is that productivity is improving, so that fewer workers, and in some cases less-skilled workers, can produce more of a better product. With the sharp increase in demand by airlines, employment is again picking up in the aerospace industry, but given the new efficiencies noted above and the continuing low demand for defense products, it is unlikely to return to former levels.

Currently about 50 percent of U.S. aerospace products is sold to the U.S. government for defense, space, and air traffic control. Of the other 50 percent, about 75 percent relates to exports of both commercial and military products. Government purchases are expected to remain flat, so that most growth in the industry will depend on success in the international marketplace. There are several factors that bear watching on the international horizon that could have an impact on the pre-eminent role of U.S. aerospace. Certainly competition in aerospace and defense products has increased, with other countries, particularly in Europe, improving the range and quality of their products in recent years. Aerospace and defense have become the glamour industries of the 1980s and 1990s, with every industrializing country attempting to stimulate some domestic aerospace and/or defense capacity. This is particularly troubling in the defense arena where there is serious surplus capacity for production of military products in Europe, Russia, and the United States at precisely a time when other nations are attempting to build their own indigenous capability.

As with other industries, there is also an increasing tendency toward internationalization or globalization of the aerospace and defense sectors. That is, U.S. companies depend for a large portion of their sales on foreign markets and increasingly have found it useful to work with foreign companies on some projects or to obtain components and technologies from off shore when it is economically advantageous to do so. In the commercial aircraft and engine sector, the enormous cost of launching a new product has increasingly led to the formation of international partnerships so as to spread risk, obtain financing on more favorable terms, improve access to markets, and to obtain the best technology available or to avoid having to develop technology that already exists. Similar considerations have led the U.S. government to encourage cooperative military projects.

Some industry observers and labor representatives have argued that this internationalization process, particularly the outsourcing of aerospace production overseas, threatens U.S. jobs and in the long run can create future competitors. These are the same arguments used against offsets. However, there is little evidence that the reduction in employment over the past few years has anything to do with increased reliance on offshore procurement of aerospace parts and components. In fact, during the period 1990–1995, imports of aerospace parts and components for aircraft and engines have accounted for a rather stable 5.5–6.0 percent of total U.S. aerospace production. What is clear is that, were it not for exports, U.S. aerospace employment would have dropped even more significantly during the 1990s.

It is against this background that the question of offsets must be examined. For purposes of this review, it is assumed that offsets refer to the various conditions of sale that a foreign government imposes on the United States and other vendors that are in addition to supplying the desired military product. Such offsets may include direct offsets related to the product sold. Examples include allowing the purchasing country to manufacture some share of the product it is purchasing and even provide parts for similar products being sold to other countries (including the U.S. government). It may include training and transfer of technology to undertake such production or similar assistance to help the customer maintain the product being purchased. An agreement may also include indirect offsets, which can cover many different activities. The selling company may purchase goods from the customer country that are unrelated to the item being sold. This is particularly attractive to companies that have large nondefense sectors that require intermediate goods from U.S. and international sources. It may involve providing marketing assistance to increase overall exports or exports from certain preferred sectors. Indirect offsets may involve investing in the purchasing country to stimulate nontraditional industry.

It should be noted that industry does not regard a straightforward licensing or co-production agreement, which does not involve the sale of a product directly from the United States, to be considered an offset in terms of this policy discussion. Although there are certainly transactions that fall in a gray area, licensing or co-production agreements to produce U.S.-designed military products, which do not involve the sale of the end item from the United States, are no more offset related than a U.S.-owned automobile line in the United Kingdom or a U.S.-owned or U.S.-licensed electronics operation in Hong Kong. Thus the Japanese production of the F-15 under license from the United States is not considered an offset program, nor is the U.S. production of the Joint Primary Aircraft Trainer System (JPATS) under license from the Swiss.

From an industry perspective, offsets are often considered a nuisance. Most companies would much prefer to compete on the basis of the quality and price of their primary product. Aerospace companies are generally not in the consulting, technology transfer, risk capital, or trading business. However, just as in the commercial aerospace arena, it has become necessary to find imaginative means to help customers finance their aircraft; in the international military market, offsets have become a recognized part of doing business with most government customers.

Offsets are of course not a new invention, but at least in part simply another form of the age-old practice of barter and countertrade. Although inefficient, it should be remembered that, for every export a country makes, mathematically at some time and from some place there must be a corresponding import, unless a country is giving away the original export. That import will negatively affect some producer, but the society as a whole will generally benefit. Offsets in part close the trading loop in a bilateral and visible fashion, but they do not change the basic principles of trade.

Furthermore, countries could generally obtain independently much of what they gain through offsets. The United States and other countries export billions of dollars of machinery every year that is used by purchasers to produce new or better products. Other companies specialize in providing customers with "turn-key" factories, tailor-made software, and consulting services for technology development, administration, and marketing. In general we applaud such exports of goods and services from the United States, even when in the long run they help create competitors overseas. In one sense, a sale with offsets is simply a way to bundle a number of goods and services in a single export package.

Finally, it should be noted that offset requirements are not unique to dealing with overseas governments. American prime contractors for defense products are required to perform a number of activities for the U.S. government that are not demanded of commercial transactions, such as setting aside business for small and minority-owned enterprises, adhering to unique cost-accounting standards, meeting military specifications that may have no relationship to commercial markets, and assuring the widest geographic spread of subcontracts and vendors consistent with meeting price and quality standards. When government is a customer, whether foreign or domestic, vendors are often expected to provide more than the best product for the best price.

Critics of offsets also need to be reminded that when the U.S. Department of Defense (DoD) makes a major purchase of a foreign-designed weapons system, it almost always demands that it be wholly or in large part produced in the United States. Recent examples include the AV-8A and B Harrier, the T-45 Goshawk, the Multiple Subscriber Equipment system, the 9-mm Beretta pistol, and the Joint Primary Aircraft Trainer System. Although the United States regards this "domestic production line" requirement as related to security, to U.S. foreign trading partners it looks very much like a 100 percent direct offset policy.

One other caveat needs mentioning. As the United States has the world's largest economy, it can be argued that offsets can provide a form of marketing advantage to U.S. firms. That is, the United States can absorb offset requirements, including some purchases from the customer country, with little or no impact on the overall U.S. economy more readily than can U.S. competitors. This marketing tool is particularly important to the U.S. defense industry given the lack of U.S. export finance for defense products and technology transfer controls that often preclude the United States competing with its best technology. The capacity of the U.S. economy to absorb offsets must be weighed against possible costs to the economy.

In 1995 the U.S. gross national product amounted to $7 trillion. Total imports equaled $743 billion. Total aerospace production totaled $106 billion. By contrast, the most recent U.S. Department of Commerce study of offsets indicates that in 1995 total offset transactions for which prime contractors received offset credits totaled $2.7 billion.[1] Recognizing that some of the offset values are negotiated credit figures that are often higher than actual market prices and that a number of these transactions involved switching from one foreign source to another rather than involving any potential U.S. sale, this number is clearly an outer boundary relative to a comparison with the general U.S. economic data. Of those transactions, about 60 percent were indirect offsets, thus involving sectors outside the defense sector. On an aggregate basis, it is hard to imagine that $2.7 billion in offset transactions would have much of an impact on an economy the size of the United States.

After over ten years of studies of offsets by the executive branch and the U.S. General Accounting Office, no clear evidence has materialized that offsets have had any significant negative impact on specific sectors or subsectors of the U.S.

[1]Department of Commerce, Bureau of Export Administration. 1997. Offsets in Defense Trade.

economy. A variety of hearings over recent years by Congress and the International Trade Commission have had witnesses allege damage related to offsets, but provided no specific examples of such damage. The 1997 Commerce Department study on offsets focuses on three industrial base subsectors that have seen their U.S. market share deteriorate over the years—gears, machine tools, and shipbuilding.[2] In each case offset performance for the period 1993–1995 was less than 1 percent of domestic production of the sector during the same period. It was also not clear to what extent offset credits were generated for purchases that were diverted from one foreign supplier to another, rather than from a potential U.S. supplier.

There are perhaps occasions when both the U.S. government and the prime contractors have not exerted enough caution to ensure that they do not establish a climate in which subcontractors can be unduly pressured by foreign customers into agreeing to licensed production as part of an offset or co-production agreement. Improved communication among DoD officials, primes, and subcontractors could help avoid such situations.

There are certainly problems related to the defense industrial base. But it is likely that offsets tend to reflect those problems, not cause them. Overall, the U.S. defense industry suffers far more from other problems—for example, the recent sharp decline in the DoD procurement budget, DoD acquisition regulations, high costs of venture capital, and a financial system geared to short-term returns rather than to long-term improvements in productivity and product—than it does from foreign-imposed offset requirements.

In 1990 the Aerospace Industries Association submitted a paper to the executive branch suggesting that it incorporate a set of principles into any federal policy with respect to offsets. These principles, outlined below, still represent the thinking of the major aerospace and defense companies:

1. The U.S. government should not take unilateral measures through statute or regulation to control offsets that would simply cause business to go to foreign competitors (except for current technology transfer restrictions related to security).

Comment: Most U.S. products that are eligible for export must compete with similar equipment produced in other countries. It does no good for U.S. producers or the industrial base if the United States restricts offset offers by U.S. firms that simply result in a customer turning to another supplier who is willing to provide the equipment and a satisfactory offset package. Industry has seen this happen with great frequency in the unilateral application of foreign policy and national security export controls. Almost invariably the end result is to shift demand to other suppliers without obtaining the desired foreign policy or security objective. The government should avoid yet another unilateral form of controls that is even more market-distorting than the practice it was set out to discourage.

[2]Op.cit., pp. 54-59.

2. Direct offsets should not be allowed when a purchase is wholly financed by U.S. assistance on grant terms, except when there is agreement by DoD and U.S. defense firms competing for the business.

Comment: When U.S. grant funds are provided to a country, it is generally required to use them to purchase U.S. products. Hence a unilateral limitation on offsets by the U.S. government is unlikely to result in a country refusing to buy from the United States. It is true, however, that a country contemplating the purchase of more than one system might well be influenced to purchase one or another from the United States depending on the offset offers made by other countries. Hence, unilateral controls might favor one product and one U.S. firm over another, but the total purchases from the United States are unlikely to be affected.

As noted above, there may be some cases in which the U.S. government might determine that it is to the U.S. advantage to allow a U.S. ally to use grant funding to establish some domestic production capacity for a U.S. product, as in the Egyptian MIA1 tank sale. In such cases, if agreement is reached with the U.S. defense firms involved, limited funds might be used for such purposes. However, such use of funds should be avoided if at all possible, as it risks undercutting support for the overall military assistance program.

3. Efforts should be made by the United States to obtain a multilateral agreement on disciplining offset practices, or at least to obtain understandings with the major U.S. defense trading partners to restrain their offset demands.

Comment: Ideally, some international code of conduct on offsets might be negotiated, which would reduce or eliminate offset demands without prejudicing U.S. suppliers. Examples of similar agreements include the Arrangement on Guidelines for Officially Supported Export Credits that was negotiated in the Organization for Economic Cooperation and Development, and the World Trade Organization Agreement on Government Procurement. Industry is not overly optimistic on this score. In the first place, the effort to limit official export credits has been only moderately successful, with foreign governments quickly finding ways around the agreement to help their own firms (e.g., mixed credits). In a world in which most foreign governments purchase far more civilian goods than the U.S. government (e.g., surface and air transport, communications, power generation, etc.), the room for back door offsets is quite large. It would be difficult to prove that there was a formal relationship between one country's purchase of a European military product and a purchase by that country's national telecommunications company from the customer country. Indeed, it might be difficult to determine when a U.S. aerospace company's purchases from a given country were or were not related to a purchase by that country of the U.S. company's military products.

Furthermore, in any negotiations on offsets U.S. allies are almost certain to demand that, if they are to limit offset demands, the United States must be willing to negotiate on some of its procurement practices. These would include its "buy America" laws, its domestic set-aside programs that do not allow foreign participation, and the general U.S. insistence on a warm production line for any major system it purchases from off shore. As it is unlikely that the United States will be politically or militarily willing to do so, negotiations are unlikely to be very successful.

It is perhaps more probable that the inefficiencies and political irritation caused by offsets might eventually lead the major U.S. allies, such as Canada, Australia, Korea, Britain, and France, to at least reduce their offset demands, either explicitly or informally. This is unlikely to happen, however, until Europe succeeds in further downsizing and rationalizing its defense industry to more closely match current demand. We assume that U.S. negotiators will explore such options.

4. In cases in which the only competitors for a foreign contract are U.S. firms, the U.S. government might play a useful role in limiting offsets, but should do so only after full consultations with the U.S. firms involved in the competition.

Comment: As a general rule, industry believes that the government should support all U.S. companies in their efforts to compete against foreign companies, but should not attempt to intervene in individual company offers with respect to price, terms, or content of offers, including offsets (except for security-related technology controls). However, there are rare cases in which U.S. companies appear not to face foreign competition. In these few cases in which either the foreign government must buy a U.S. product for political reasons, or because there is simply no other comparable product available, the U.S. government might consult with U.S. companies involved to determine the feasibility of placing some restrictions on offset offers. This could prevent two U.S. companies from escalating offset offers, or even a single U.S. company from being pressured by the foreign government into making excessive offers.

There are three major dangers to such government involvement. First, any formal U.S. government policy of intervention when there are no foreign competitors might simply encourage the foreign customer to stimulate such competition, to pursue a different approach to addressing their defense problem, or to decide against any purchase. Second, the U.S. government might bring undue pressure on U.S. companies to agree to government involvement, even when such interference might favor one company over another. Finally, if confronted with a percentage cap on offsets, a country might well demand higher quality offsets from companies, which might be more onerous than a higher percentage.

5. The collection and publication of information on offsets by the government should be handled with extreme caution. Such information, particularly when attempts are made to standardize the data, can be very misleading and thereby exaggerate the U.S. perception of the problem, encourage even stronger

demands by other governments, disclose proprietary information, and possibly damage the competitive position of U.S. firms in international defense business.

Comment: U.S. industry has generally opposed massive data collections on offsets. It has done so because each offset is so unique—and the meaning of individual numbers (particularly percentages) so particular to a specific offset program—that aggregating the data may obfuscate the issue more than it clarifies. It should be noted that there is a clear incentive for the selling firm and even some agencies in the purchasing government to inflate the offset figures so as to put the best image possible on the purchase of a foreign product. However, this in turn tends to distort the importance of offsets in U.S. studies. It also may well escalate the demands of other countries that read the reports.

In general, if the concern is over the impact of offsets on the defense industrial base, the government would be better served by identifying which specific industries seem to be in trouble. Studies of those industries should then identify all the sources of their difficulty, including offsets. It is likely that if such studies are conducted, it will be found that offsets are not a very important aspect of problems relating to the industrial base. It may well be the case that even where offsets appear to affect a weak subsector of the economy, it will turn out that offsets are more a symptom than a cause of the problem. That is, subsectors that have not kept up technologically, or employ technologies that are widely available around the world, or that are employing high-paid labor in low-skilled jobs are precisely the types of sectors for which foreign alternatives prove attractive, with or without offset obligations. However, if government can find specific subsectors that are particularly impacted by offsets, it would then be possible to work with the limited number of companies that account for most offset performance to see if greater restraint could be used in meeting offset obligations in such subsectors.

In summary, the aerospace industry has concluded that, whereas the U.S. government might continue to make clear its dislike for offsets to U.S. trading partners, it should use extreme caution in taking any action that would simply shift purchasers away from U.S. producers. The government essentially agreed with those views, and in 1990 the White House issued a policy statement[3] on offsets that tracked closely with the industry recommendations enumerated above. That statement remains the policy of the U.S. government, which industry supports. Although all policies must change with external circumstances, so far there do not appear to be any developments in the offset world that would justify a departure from current policy. The aerospace industry certainly is prepared to continue to work with the government to monitor the offset landscape to ensure that current policy continues to be appropriate.

---

[3]Statement by the White House Press Secretary, April 16, 1990.

Dual-Use Supplier Management and Strategic International Sourcing in Aircraft Manufacturing

Todd A. Watkins
Lehigh University

INTRODUCTION

Supplier development and management systems of U.S. aircraft manufacturers have evolved rapidly during the 1990s in an environment of reduced military procurement and a worldwide slump in commercial aircraft sales. The result has been an ongoing restructuring within the aircraft industry supplier base, with significant implications for customer–supplier relationships and, potentially, national security. For example, with excess capacity and increasing cost pressures, many aircraft firms are dramatically reducing their numbers of suppliers.[1] At the same time, many of these suppliers are being asked to take on additional responsibilities in design, assembly, materiel management, risk sharing, and even in fostering the internationalization of their customers' supply base.

Prior research on such "lean" manufacturing principles and associated supplier management practices relied largely on the study of the automobile industry (Womack et al., 1990; Womack and Jones, 1994; Helper, 1991; Nishiguchi, 1994). This case study of the anonymous Generic Aircraft Manufacturing Company (GAMC) is part of a larger ongoing effort, through the Massachusetts Institute of Technology (MIT) Lean Aerospace Initiative, to extend the study of lean manufacturing practices to the aircraft industry and to evaluate their applicability.

Indeed, GAMC is very actively engaged in implementing total quality man-

[1]For example, unpublished data show that the average number of suppliers projected for 1995 by the business units in the survey were more than 50 percent below 1989 levels. This from the Massachusetts Institute of Technology Lean Aerospace Initiative Supplier Systems and Relationships Survey of nearly 80 business units of the U.S. aircraft industry's top manufacturers.

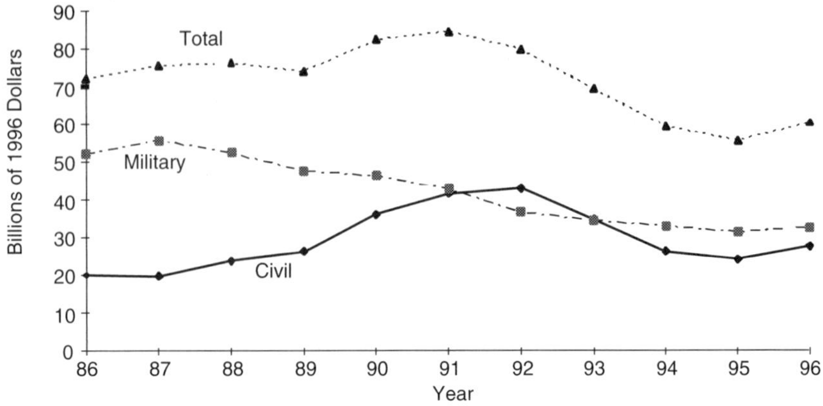

FIGURE 1 U.S. aircraft industry sales, 1986-1996. Source: AIA (1994 and 1995).

agement techniques and the entire lean–agile manufacturing paradigm, as it understands these concepts and practices. For example, to name a few techniques GAMC has adopted, they have implemented an intensive statistical process control program throughout the firm; developed and begun using a supplier rating and certification system; restructured the entire company into integrated, multifunctional product–process development teams; participated in early supplier involvement programs of their primary customers; involved their suppliers in similar efforts; entered into teaming and long-term risk-revenue sharing partnerships with customers and suppliers.

The aircraft industry differs substantially from the automobile industry. Thus, a major research goal is to understand the implications of those differences. The choice of GAMC as a target for a first exploratory case analysis is not intended to identify or suggest "best practice" or "poor practice" in the aircraft industry. No stand-alone case can claim such insight. Rather, the case investigates the implications of three major differences between the auto and aircraft industries.

First, in automobiles the lean manufacturing system was pioneered by Toyota, and emulated by other Japanese manufacturers, during a postwar period of remarkable economic expansion. By comparison, as Figure 1 shows, the U.S. aircraft industry in the 1990s saw a large-scale downturn. The Aerospace Industries Association (AIA) estimates that 1995 U.S. aircraft sales were $54 billion, down 34 percent in real terms from their 1991 peak (AIA, 1995). U.S. aircraft manufacturers have, for the most part, only within the last few years begun implementing techniques drawn from research and writing on lean manufacturing. Clearly techniques for dealing with suppliers could be very different between rapidly expanding and rapidly shrinking industries.

A second major difference is the commercial–military split in the U.S. aircraft industry. As Figure 1 also shows, at the 1987 peak of the U.S. defense buildup, nearly 74 percent of the aircraft industry's output was defense related. Defense production still accounted for more than half in 1996. One key area for research, then, is understanding how firms develop supplier networks across the two sides of the industry. In particular, how can firms maintain cost-competitive supplier networks for meeting the demands of both military and commercial customers, and what are the implications for supplier management strategies?

A third major difference with the automobile industry is the sheer complexity of today's aircraft. A typical modern automobile has about 10,000 parts (Womack et al., 1990). The new Boeing 777 has 4 million. As a result, the final aircraft assemblers such as Boeing and Airbus rely on suppliers to build and integrate considerably complex assemblies (such as wings complete with hydraulics, fuel monitoring systems, flaps, and thousands of rivets). Major aircraft subassemblies arguably are as complex as entire automobiles. Supplier management in the aircraft industry therefore takes on additional importance in terms of cost competitiveness and the quality and performance of the final product.

Moreover, the first-tier aircraft industry suppliers such as GAMC not only have to respond (looking upward in the supply chain) to the supplier management efforts of their major customers, but also must (looking downward) manage extensive supplier networks of their own. Womack and colleagues (1990) compare the 340 production suppliers that the "lean" Japanese automobile manufacturers typically have to the 2,500 used by General Motors. As indicated below, GAMC manages a supplier network of comparable size.

GAMC AS MICROCOSM OF THE INDUSTRY

GAMC produces major aircraft structural sections for both commercial and defense-related customers. In 1994 GAMC was acquired by a new parent firm, itself created by a recent merger. As of early 1995, and before restructuring under new ownership, GAMC employed approximately 5,000 people and had annual revenues near $600 million. GAMC was made the home of the new parent company's commercial operations.

Both before and after the acquisition, GAMC has seen itself as a major subcontractor with a focus on integration of major structural assemblies, such as wings, tails, and engine nacelles. GAMC's strategic vision has been that of moving more toward full responsibility for complex design and integration and gradually away from in-house detailed part fabrication.

The recent acquisition is but the latest episode in the long history of the company. Until the 1980s, GAMC for many decades had been a prime contractor producing complete military aircraft as well as major structural assemblies for commercial aircraft. After the Vietnam War, however, defense prime contract work fell, and the company moved into integrating large structural assemblies for

TABLE 1 Major Reductions and Restructuring at GAMC, 1991–1995

	1991	1995[a]
Sales ($ millions)	1,000	600 (est.)
Employment	9,700	5,000
Approved suppliers	1,400	725
Civil percent of sales (approx.)	40	60
Owner	Diversified conglomerate not focused on aerospace	Top-tier aerospace and defense company

[a]1995 figures do not include restructuring with new parent firm.

the top-tier final assemblers in both the military and the civil segments of the industry.

GAMC in many ways reflects the aircraft industry as a whole. It has been through several major restructuring efforts and changes in ownership over the past decade, in large part due to the continuing consolidation in the U.S. aircraft industry as a whole. As Table 1 shows, changes since 1991 at GAMC mirror the downsizing and restructuring of the aircraft industry more broadly. GAMC's sales fell about 40 percent between 1991 and 1995, from more than $1 billion per year. Employment fell to about 5,000 from 9,700. Floor space, at one time more than 8 million square feet, also contracted by about 40 percent to 5 million square feet. Throughout this downsizing process, GAMC also underwent significant restructuring, including several different owners. Like nearly every upper-tier firm in the industry, GAMC also significantly reduced its number of approved production suppliers, from more than 1,400 in 1991 to 725 in 1995. And, like the industry as a whole, while GAMC retains a balance of commercial and military production, the commercial fraction increased. GAMC managers report that the mix shifted to about 60 percent nondefense work in 1995 from about 40 percent in 1991.

GAMC is also of interest for study as a first-tier supplier to the major aircraft final assemblers (the "airframers"). GAMC's strategy has been to place itself immediately below the major airframers in the supply chain and to increase its own responsibility for integrating major structural aircraft assemblies and managing subcontractors for these top-tier firms. Thus, GAMC's preacquisition pro-

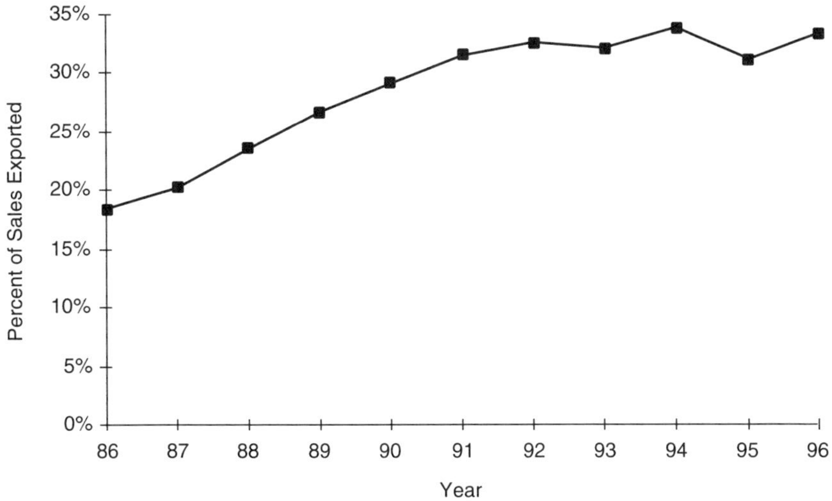

FIGURE 2 U.S. aerospace exports as a percent of total sales, 1986–1996. Source: AIA (1994–1995).

motional materials described GAMC's market niche as a "support partner" and a "major subcontractor with prime capabilities: integrated concurrent engineering, sophisticated testing labs, and advanced manufacturing." GAMC thus provides an opportunity to examine the effects in the supply base of the supplier management practices of the top-tier aircraft companies.

In addition, as a supplier of complex integrated assemblies, GAMC itself must manage a network of more than 700 lower-tier suppliers of subassemblies, detail parts, and raw materials. Given a supplier network comparable in magnitude to those of major auto suppliers, GAMC's supplier management practices are of interest not only as a significant challenge on their own, but also because of how they are shaped by GAMC's customers. Like its customers, GAMC is delegating increased responsibilities to its own suppliers for design, quality control, risk sharing, and supplier management. This reflects an apparent trend in the aircraft industry more generally, following the lead of the Japanese automobile industry, toward turning historically loosely tiered and arms-length subcontracting relationships into more tightly tiered, more closely controlled structures.

At the same time, GAMC's supplier management organization has had to adapt over the past decade to the globalization of aircraft markets. As Figure 2 shows for the U.S. aerospace industry, of which aircraft is by far the largest segment, exports have nearly doubled as a fraction of total sales over the past decade. GAMC's major customers, the major players in aircraft export markets, are seeking growth globally and are aggressively turning to GAMC for help.

In sum, both GAMC and the U.S. aircraft industry are rapidly moving targets. Caution is clearly in order concerning interpretation of conclusions from this (or any other) single case. Nevertheless, GAMC does represent a reasonable microcosm of the aircraft industry as a whole. It therefore provides considerable insight into the implications of major differences with the automobile industry as well as into some of the major issues and tensions confronting supplier management in the U.S. aircraft industry today.

In particular, the discussion below focuses mainly on three related issues: (1) GAMC's management of suppliers to both its military and commercial programs with a single dual-use integrated materiel and procurement system; (2) the impact on GAMC's supplier management practices of the globalization of its customers' markets and strategies; and (3) the tensions created by GAMC's implementation of "lean" supplier management practices at the same time that the industry is restructuring and globalizing.

DUAL-USE INTEGRATED SUPPLIER MANAGEMENT

How did these major restructuring efforts affect GAMC's supplier management and materiel operations? These operations were somewhat insulated from the organizational upheaval over the decade prior to the acquisition because throughout the period these functions remained integrated across the whole organization. Except for international strategic sourcing, the changes discussed below were related largely to the implementation of practices suggested by "lean" manufacturing principles.

Before the acquisition, a common functional group with a single procurement system supported all GAMC programs and divisions (whether commercial or military). Approximately 400 people at GAMC performed materiel functions, including subcontract management, supplier development and technical support, inventory control, purchasing, administration, procurement quality, receiving, warehousing, shipping, and off-site personnel. Purchases from suppliers and subcontractors represented an estimated 35 percent of GAMC's cost of sales.

GAMC's main decision-making tool concerning suppliers was its "preferred supplier process." This supplier management system included supplier assessment, certification, and selection systems; technical assistance to suppliers to foster supplier statistical process control (SPC) and total quality management (TQM) capabilities; passing more responsibilities to sub-tiers for design, quality, risk sharing, and management of lower tiers; and longer-term teaming where possible. These efforts were relatively new to the company, having been implemented beginning in the early 1990s.

The supplier assessment system included GAMC's "supplier rating system" (SRS) that collected data on supplier performance to quantity, schedule, and documentation. Essentially, SRS scored suppliers on how well they met the terms of purchase orders. The next level of GAMC's supplier assessment system, the "supplier performance improvement program" (SPIP), added a metric to track quality: the cost of defects discovered after receipt by GAMC.

The certification program, called the "Qualification System," applied these supplier assessments to determine which suppliers were approved for doing business with GAMC. When a supplier's performance score fell outside acceptable limits, the supplier was removed from the list of suppliers approved for bidding on new business. These suppliers submitted corrective action plans and were given an opportunity to demonstrate improvement. If the supplier remained substandard, it lost its qualified status. All suppliers, whether for military or commercial programs, were measured and certified under the same set of criteria.

In monitoring and selecting suppliers, GAMC's system distinguished three types of suppliers. First, "vendors" supplied catalog items, office and maintenance supplies. The SRS (performance to purchase order) applied to vendors. Companies typically have large numbers of this type of nonproduction (often called "indirect") supplier. Yet many, even those with SRSs, do not track vendor performance.

Second, GAMC called "suppliers" those firms with "build-to-print" capabilities, in other words those who do detail parts or assemblies and subcontract labor based on complete design drawings (prints) provided by GAMC. Both the SRS and the SPIP applied to suppliers.

Third, "subcontractors" had "build-to-spec." capabilities, did major structural assemblies, and/or had proprietary products or processes. They could be integrated with GAMC's design teams and share risks for major aircraft sections. When selecting subcontractors, GAMC managers used the SRS and SPIP for guidance, but also considered broader strategic partnering or marketing goals and the supplier's capability for greater responsibilities in design, risk sharing, and lower-tier supplier management. Examples of suppliers selected with these broader strategic goals in mind are discussed below. Special cross-functional selection teams, called "buy-to-build package teams," with members from various functional areas, were formed for decisions on subcontractors for fracture critical parts, complex machined parts, assemblies, parts with high reject histories, major sources of costs, and hot/superplastic- or spin-formed parts.

Indeed, such cross-functional teams remain the organizational norm at GAMC, making GAMC a truly dual-use company. Integrated, centralized functional groups such as engineering, machining and fabrication, quality assurance, and so on, serve all programs, both military and commercial, with the same people and procedures. The company operates under an "integrated product–process development" (IPPD) philosophy with what GAMC human resource managers call a "strong matrix" organizational structure. In addition to reporting to a functional group, one axis of the matrix, people also report to (and are collocated with) multifunctional product or process teams, the other axis, that have full responsibility for integrating and managing all aspects and the whole life cycle of each program, from development through delivery and post-production support.

Functional groups such as materials, business operations, and human resources assign members to "integrated product teams." However, members also retain ties to the functional department, importantly because responsibility for managing and improving functional processes (e.g., deep-pocket machining, materials tracking) falls to "process management teams" within the functional departments. GAMC managers expect this matrix approach to maintain communication and sharing of expertise among functional professionals at the same time as fostering interfunctional problem solving within the product teams. Accountability is shared among team members, and budgets, schedules, and decision making are allocated to teams, not departments. This integrated team organizational structure as of 1995 is shown in Figure 3, and an example team is shown in Figure 4. On the charts, IPT stands for integrated product team and PMT for process management team.

Consistent with the IPPD philosophy, GAMC's supplier management approach is to deal the same way and with the same procurement system with suppliers to both its commercial and its military programs. But the integrated approach to suppliers is also driven by the dual-use nature of GAMC's supplier base. More than 80 percent of GAMC's suppliers serve both civil and military programs. For example, aluminum alloys from one raw material supplier are milled for commercial and military programs alike in GAMC's central fabrication operations by huge robotic machining stations from another single supplier. A small California supplier does heat treat processing in the same facility for GAMC's military and commercial programs alike. Other single suppliers provide film adhesives, glass and other "pre-preg" fabrics, graphite composites, and primers for use on several different programs on both sides of the business. There is not 100 percent supplier overlap between military and commercial programs in part because occasionally program (or customer) specific requirements arise, either on the commercial or defense side, and in part because many of GAMC's suppliers must be approved by GAMC's customers, who historically have not had identical approved supplier lists.

One question that arises is whether the substantially dual-use nature of GAMC's supplier base might somehow be uniquely encouraged by the IPPD philosophy. Two research findings, however, suggest otherwise. First, there is, to a very considerable extent, overlap and integration between the broader U.S. manufacturing base serving military programs and that serving commercial markets. In a survey-based statistical study of manufacturing plants in 21 different durable goods industries (including but not only aircraft) that account for half of all durable goods purchased for defense, Kelley and Watkins (1995) found that in 1990 the vast majority of defense contractors and subcontractors served both defense and commercial customers. Indeed, as Figure 5 shows, most plants doing defense work in the industries studied, both prime contractors and subcontractors, actually do more commercial work than defense work. It is, then, not sur-

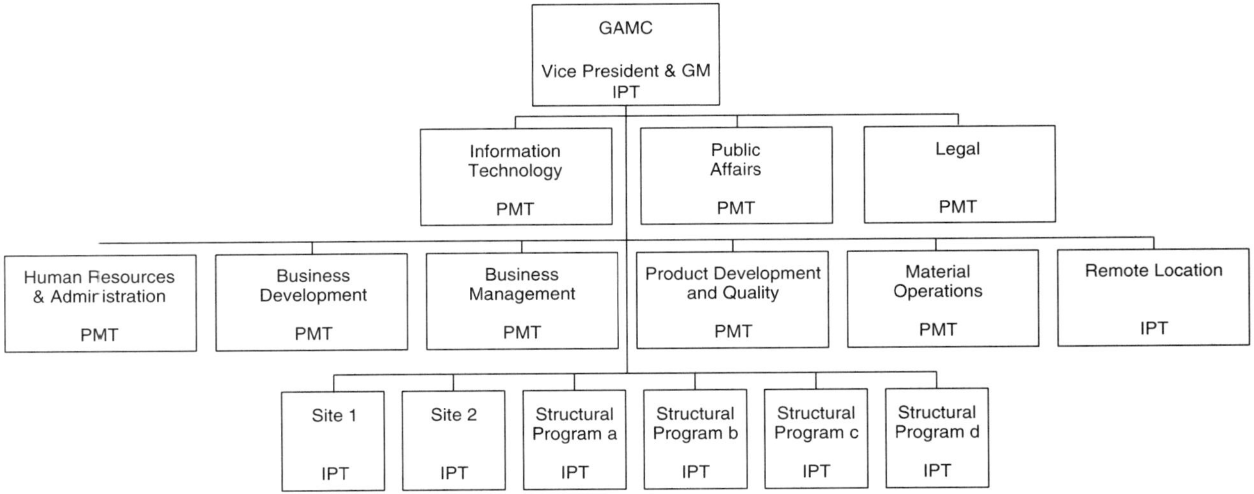

FIGURE 3 GAMC Integrated Product/Process Development organization.

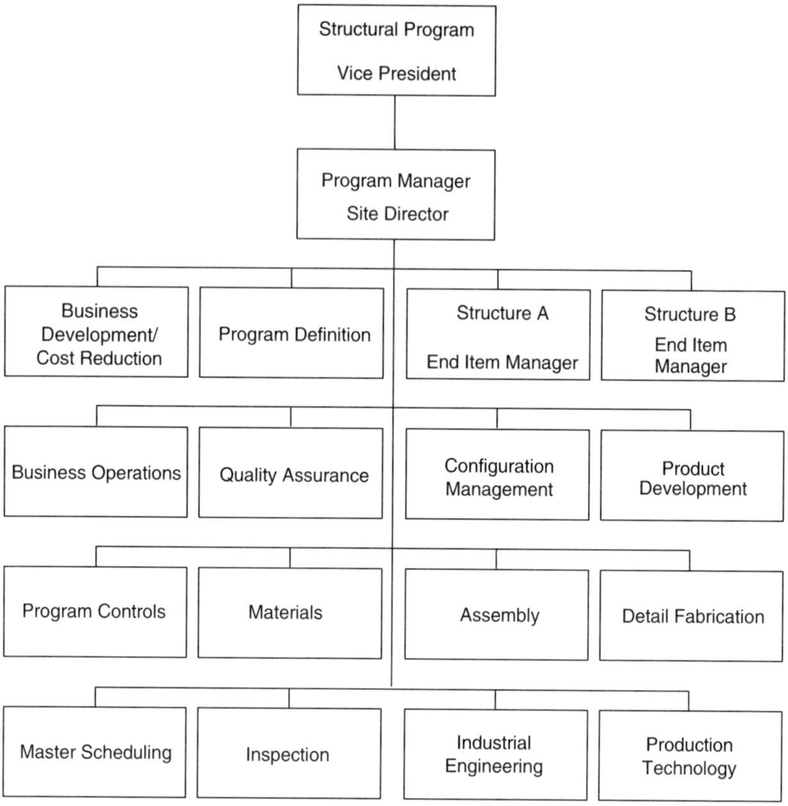

FIGURE 4 GAMC structural program Integrated Product Team (IPT).

prising that GAMC's suppliers reflect the dual-use nature of the broader supplier base.

A second indication that an integrated, dual-use approach to supplier management is not unique to the IPPD philosophy is that GAMC also used the integrated approach previously under two otherwise very different organizational structures. First, during the peak of the defense build-up in the 1980s, GAMC had separate divisions for military aircraft, for commercial aircraft, and (much smaller) for aircraft modernization and support (Figure 6).

Under this divisional structure, according to managers with the company at the time, GAMC operated basically segregated military and commercial businesses, from engineering and accounting through marketing and manufacturing. For example, the military division had a manufacturing engineering group, and the commercial division had its own manufacturing engineering group. Even the

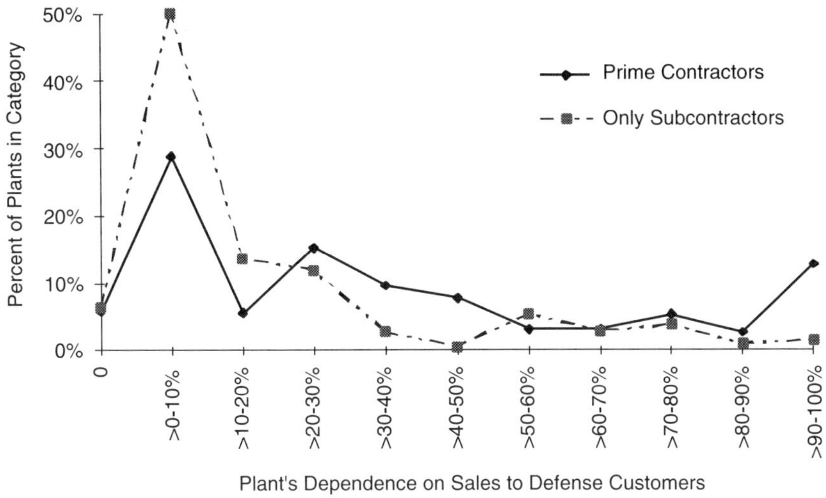

FIGURE 5 Defense contractors' dependency on defense-related sales, 1990. Source: Kelley and Watkins (1995).

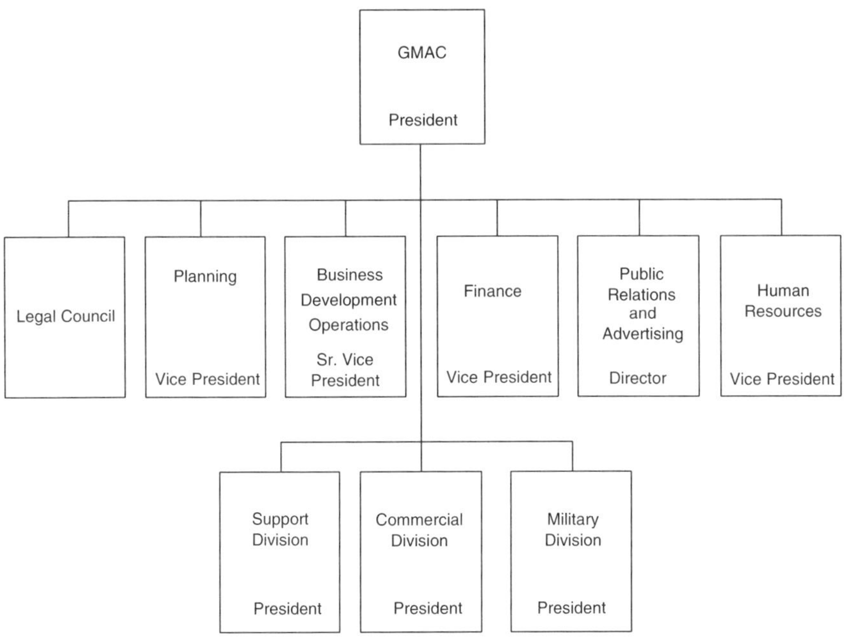

FIGURE 6 GAMC organizational structure, mid-1980s.

small aircraft modernization and support division had a manufacturing engineering group. One production manager described it as "a dividing line between the two where we could not interface. . . . It was like two totally different companies." Nevertheless, materiel functions were centralized. The same people, using a single system of procedures, served the whole, including both the military and the commercial divisions.

Then, second, at the end of the decade, GAMC reorganized from substantially separate defense and military divisions, to substantially integrated functional groups serving individual assembly programs. A major motive was the downturn in business that was due to the end of the Cold War and the rapid decline of commercial sales as foreign competition increased and airlines, struggling with cash flow problems or bankruptcy, cut back on orders for new aircraft. As business on both sides shrank, redundancies and overcapacity in functional areas duplicated across the two sides became more problematic. The organization after this (what GAMC managers refer to as) "de-divisionalization" is shown in Figure 7.

The company moved to what might be called a "weak" matrix organization. One axis of the matrix structure was divided by structural program for program management. The other axis was divided by function: including design, manufacturing engineering, logistics, central detail fabrication, and quality assurance. The centralized functional groups supported both military and commercial products. A single vice president oversaw all assembly operations, both commercial and military. The matrix was "weak" in the sense that the reporting, promotion, and reward structure relied primarily on one axis, these functional homes.

Again, supplier management, procurement, and materiel functions were common across all programs. But now this was more in line with the rest of the organizational structure. Indeed, there was near unanimous agreement across interviews with people who were performing supplier management, procurement, and materiel functions at the time that the de-divisionalization had little or no impact on how they operated. This despite the rest of the organization going from substantially separate military and commercial operations to integrated, dual-use functional organizations. To the materiel people, the transition was transparent because they had already been operating a system common to both sides of the business.

Then, by early 1995, the company had integrated further with the IPPD structure. One manufacturing manager put it this way:

> Now we're trying to break down the traditional barriers between manufacturing and engineering where you had a VP of manufacturing, a VP of engineering and a VP of materials. We've gone to where we just have one VP of operations, and he has design engineers, manufacturing, everything.

Rather than being a difficult transition for the supplier management, materiel, and procurement operations, the move actually made their integrated, dual-use ap-

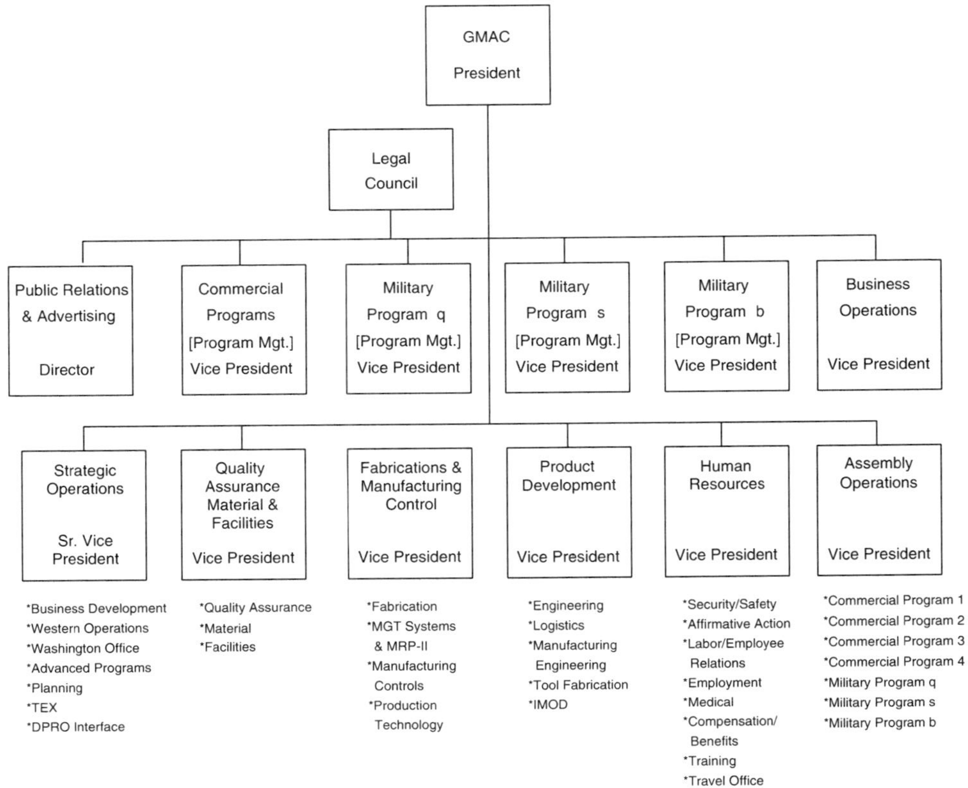

FIGURE 7 GAMC organizational structure, 1990

proach fit more neatly with the strategic philosophy of the whole organization. As a management tool, integrated product and process development teaming is inherently dual-use in a company with large shares of its business in both commercial and military markets. The main point is to improve communication and learning along both axes in the matrix organization: across different programs, whether military or commercial, and among different functional areas. The functional homes within the company take on the role of establishing procedures and best practice guidelines common to all programs, as well as helping the functional professionals learn from one another, regardless of the program to which they are assigned. The multifunctional product teams aim to encourage experts across disciplines to learn from and help one another.

GAMC's dual-use approach to suppliers has, indeed, fostered cross-program learning. By using a single set of suppliers for both military and commercial programs, learning and supplier development on one side of the business can more readily be applied to the other.

For example, GAMC worked closely with the dual-use suppliers mentioned above, their aluminum supplier, and the small California-based heat treatment process supplier to reduce distortions in heat-treated aluminum parts. The original efforts were to improve part quality and consistency on one of GAMC's military products. As one engineer put it:

> We'd get a lot of material warpage. Particularly when you got a forging they've done a lot of pounding on. Our machinists here would machine them and they'd warp and bend in the machining process. We've worked with [the aluminum supplier] a lot, and they deliver [the aluminum] in an annealed condition. We worked with another vendor . . . out in California that does a real good heat-treat job for us. It's called a high glycol quench that really controls the distortion. They'll heat treat it to the T-73, and then we'll bring it back here and do the final machining on it.

The aluminum alloy (7050) for which this process was developed was different from the alloy (7175) that GAMC used in some commercial work, in which distortion induced by the heat-treating process was also problematic. Because of their success with controlling the quality of parts using the new process on the military program, GAMC then worked with its commercial customer to apply the lessons to GAMC's commercial programs:

> We actually asked [our customer] to change the material on that. We went from a 7175 to a 7050, which is a big deal because you have different properties. So we were able to integrate that, do a lot of working with our customer and supplier. There were meetings where we had the whole chain involved, our materials people, the suppliers and [the customer's] people would be sitting in the same room, or conference calls, or phone calls. Because our suppliers have to be certified [by the customer] too, we had to make sure that [the heat-treat supplier] had that relationship with [the customer]. We just can't send our fracture critical forgings out to anybody.

TABLE 2 Advantages of Integrated Supplier Systems

1. Flexible to organizational form:
   - at GAMC used under three very different organizational structures
   - can integrate supplier systems without integrating whole company
2. Strategic fit:
   - Integrated Product/Process Development philosophy
   - "commercial practice" in defense procurement
3. Cross-program learning with suppliers and customers
4. No duplication of:
   - suppliers ("work with the best")
   - materiel/procurement systems or positions
   - quality systems
5. Structure and control imposed by MilQ-approved system:
   - "good practice" for commercial programs, both in-house and externally
   - financial information helpful in value engineering and target pricing
   - movement in "cooperative supplier relations" to more information exchange

PROS AND CONS OF INTEGRATED SUPPLIER MANAGEMENT

Based on GAMC's experience, then, there appear to be several advantages of dual-use integrated approach to supplier management, as summarized in Table 2. The first is flexibility with respect to organizational form. The integrated approach was flexible enough to have been used without much change in three very different organizational structures. Supplier management and materiel and procurement operations were integrated across the company, even when the rest of the organization was not. Second, the dual-use approach is strategically compatible with the IPPD philosophy espoused by theories and research on lean enterprises (e.g., Womack and Jones, 1994). It is also compatible with the recent movement by the Pentagon to embrace "commercial practice" in defense procurement. Third, it facilitates cross-program learning and the sharing of ideas and information between military and commercial programs. Customers and suppliers on both sides of the business benefit, as the heat treat example demonstrates. This is exactly what the IPPD matrix teaming approach attempts to encourage, by integrating functional areas across programs.

Fourth, a single materiel system for both military and commercial programs allows a single set of suppliers to serve both sides. This eliminates the need for duplication of suppliers and is consistent with the general trend in many industries to reduce the overall number of suppliers a company must manage. It also means no duplication of materiel procedures, systems, or positions within the company. Finally, it allows GAMC to maintain only one internal quality system and a single supplier quality system. Commercial and defense suppliers are not treated differently, and indeed are the same in the vast majority of cases. Thus, GAMC's supplier development efforts, GAMC's supplier quality monitoring system, GAMC buyers, and so on can serve to improve both sides of the business.

The fifth advantage expressed by GAMC managers is related to the requirements of military contracting. GAMC's procurement system must remain approved under the so-called "MilQ" regulations, which specify the kinds of information GAMC must collect and make available to the government. There was general consensus with the prevailing conventional wisdom that these regulations go too far in, as one materiel manager put it, "crossing and dotting all the i's and t's." Yet it also had some advantages managers liked for their commercial programs as well. Primarily, some managers thought the structure and control imposed by the regulations forced both GAMC and its suppliers to maintain, if not best practice, at least good managerial practice in terms of accounting and other information systems, tracking materials, negotiating and communicating with suppliers, and so forth.

For example, sharing sensitive financial information is a relatively new but increasing practice among U.S. companies adopting principles of cooperative supplier relations and joint problem solving (Helper, 1991; Lyons et al., 1990).[2] Yet suppliers providing detailed cost information has been standard practice for many years under military contracting regulations. GAMC managers report they have found this a valuable information source, particularly in practices suggested by lean theories, such as long-term contracting, target pricing, value engineering and value analysis, as well as in identifying opportunities for working with suppliers to reduce cost.

Yet GAMC's movement toward these same lean practices has also increased some of the tensions of trying to maintain a supplier management and materiel system integrated across defense and commercial sides of the business. A particular disadvantage of the dual-use approach has been its inflexibility with respect to "life-of-product partnering." In GAMC's newest commercial program, the company entered into a revenue sharing agreement with a top-tier commercial aircraft final assembler to design and make a major critical structural assembly for its new aircraft (call it the "COM-Z"). Indeed, as discussed in more detail below, this commercial customer has revenue sharing partners worldwide. The willingness of suppliers to take on more risk was one of the customer's major supplier selection criteria.

---

[2]The same appears to be true in the defense aircraft industry. Unpublished data from the MIT Lean Aerospace Initiative Supplier Systems and Relationships Survey of nearly 80 business units from the U.S. aircraft industry's top manufacturers, for example, show that 53 percent of surveyed business units regularly receive proprietary financial information from their major suppliers.

By undertaking design and capital investment responsibilities for this major critical structure, GAMC takes considerably more business risk than under traditional subcontracting if the COM-Z is a commercial failure. In return, GAMC has contractual guarantees to get orders from the customer for the commercial life of the product and stands to garner more return per unit should it be a success. GAMC has also, in turn, a life-of-product revenue sharing agreement with one of its Japanese COM-Z strategic source subcontractors, discussed below.

Such long-term contractual relations are increasingly common in the U.S., both within and outside the aircraft industry, and have been the norm in Japan for several decades (Helper, 1991; McMillian, 1990; Nishiguchi, 1994). By increasing the buying firm's commitment to the supplier, life-of-product contracts encourage suppliers to take on the increased up-front risks of investing in design, development, and equipment. The revenue sharing is in part designed to align the supplier's incentives for quality and cost control with the buying firm's incentives. However, these type of agreements do constrain the buying firm from sending work out for new competitive bids later.

Herein lies a major tension within a single procurement system across both military and commercial business. Defense procurement requirements call for periodic "recompeting" on subcontracts. It is, therefore, problematic on the military side of the business to make guarantees for purchases from a subcontractor for the entire life cycle of programs. GAMC's procurement system is approved and regularly reviewed under these regulations, and recompeting has been the normal practice. However, there is a clear empirical trend from commercial customers toward more "partnering" and long-term contracting. The single system is strained as GAMC moves to increase its partnering with its own suppliers on commercial programs.

There is a formal mechanism in place allowing GAMC's top materiel managers to deviate from their standard (and MilQ-approved) procurement practices when they believe commercial programs have special needs. Although commercial programs now make up more than half of their business, such deviations have occurred in less than 1 percent of buying actions. However, the new COM-Z program is only just coming into production: the first major critical structures were recently delivered. The need for deviations may increase substantially should "partnering" become the commercial norm. Maintaining a single procurement system throughout the business may therefore become problematic.

Although the outcome of ongoing procurement reform is as yet uncertain, this increasing tension may be reduced. The Pentagon and Congress are signaling significant change in procurement regulations. Countless congressional hearings have addressed the broad issues, and a number of large-scale experiments are under way (e.g., the F-22 program) allowing for, among other things, long-term partnering and best value rather than best price selection of subcontractors. Whether this will become the regulatory standard, however, remains to be seen.

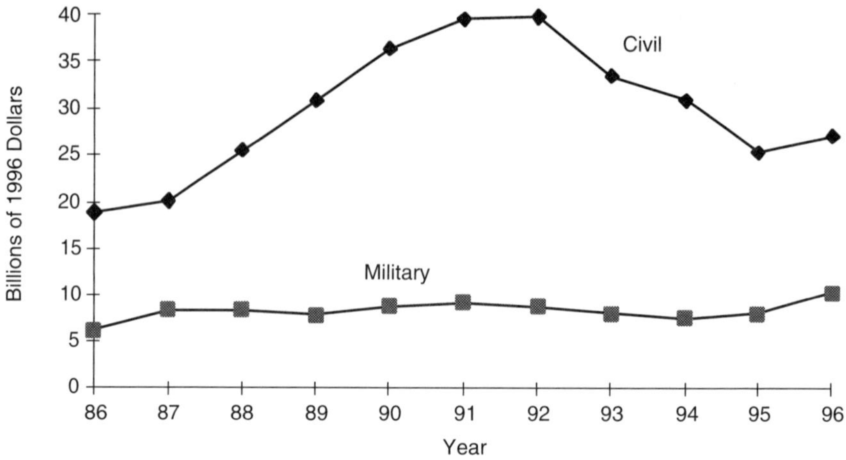

FIGURE 8 U.S. aerospace exports, 1986–1996. Source: AIA (1994 and 1995).

INTERNATIONAL STRATEGIC SOURCING

A second area of tension between commercial and defense programs in GAMC's supplier management has to do with the globalization of aircraft markets and the aerospace industry supply base. Indeed, the single major difference between GAMC's supplier management approach on the two sides of the business is in what GAMC calls international "strategic sourcing," efforts to increase its use of foreign suppliers. The purpose is to help GAMC's customers increase their competitive position in foreign markets. This program is entirely geared to support GAMC's commercial programs. It has no counterpart for military programs.

Several factors contributed to GAMC's seeking to increase its use of foreign suppliers. First, growth markets for aircraft sales have largely been overseas. Figure 1 above showed that even though U.S. defense aircraft sales began falling after 1987, industry sales as a whole continued to rise until their 1991 peak. As Figure 8 indicates, however, this growth came nearly exclusively in overseas markets. More specifically, the growth was in civil exports. Military aerospace exports are small compared with civil exports and have been flat in real terms since 1987. GAMC's international strategic sourcing efforts grew directly from the push into these overseas growth markets by its commercial customers.

Second, foreign markets are particularly important to GAMC's major commercial customers. Prior to the consolidation with its new owner, GAMC's single most important customer accounted for the vast majority of their commercial business. In the five years through the end of 1996, more than 70 percent of that customer's commercial sales were outside the U.S. Significantly for GAMC, this

TABLE 3 U.S. Civil Transport Aircraft Backlog

As of 9/30/96	Total Backlog	GAMC's Single Largest Program
Number of aircraft	1,447	X
Number that are foreign	741	More than 0.8X
Foreign as percent of total value	62.6	n.a.

SOURCE: AIA (1994 and 1995).

foreign fraction of orders was above 80 percent for its single largest commercial program. This is much higher than the already high industry average. As Table 3 shows, according to the AIA, as of the end of 1996 almost 63 percent of the U.S. civil transport aircraft industry's unfilled order backlog was from foreign customers. Foreign sales will also be critical to commercial success on GAMC's newest program, the COM-Z. The customer designed the COM-Z as a transpacific aircraft. Pacific Rim countries are obviously the key market.

Third, local content is a major factor in purchasing decisions by GAMC's customers' customers, particularly foreign government-run airlines. Thus, GAMC's customers' foreign marketing efforts entail, in part, using suppliers in key foreign markets. The pressure on GAMC for international sourcing, sometimes called "offsets," has come entirely from the commercial side of the business.

All the top-tier airframers are aggressively pushing foreign sourcing as a marketing tool. For example, an advertisement Boeing ran in a major international news magazine, *The Economist*, in early 1995 highlights the importance to Boeing of foreign sales and also of overseas suppliers: "And aerospace firms in nations around the world help build our jetliners. Trade works both ways. The more they sell us, the more we sell them." Indeed, fully 25 percent of the value of Boeing's new 777 is from Pacific Rim sources, the highest fraction of Asia–Pacific components of any previous widebody (*International Herald Tribune*, February 23, 1994). Boeing's five major Japanese partners alone have a 20 percent risk sharing stake (Boeing Company, 1990). The European Airbus consortium ran a similar ad in the same magazine several issues earlier consisting of a map of the United States showing the location of all its American suppliers. For similar local content reasons, the China National Aero-Technology Import and Export Corporations (CATIC) has co-produced about 40 McDonnell Douglas MD-80 jet transports in China (U.S. Department of Commerce, 1994). McDonnell Douglas also recently let a $1 billion contract to Halla Engineering and Heavy

Industries in Korea to supply wings for the MD-95 (*Korean Economic Daily*, November 30, 1994).

Similarly, the fourth largest civil airframer, Canada's Bombardier, in 1996 unveiled its new ultra-long-range corporate jet, the Global Express, with a marketing bash called "The Power of Global Vision." According to *Aviation Week and Space Technology*, this was

> a reference to the international team [of 9 partners across six countries] assembled by Bombardier to develop the Global Express. It was unveiled to the accompaniment of "a 45-piece orchestra [and] a 1,000-person choir." The suspense mounted with the appearance of images of the first aircraft and the choreographed appearance of the flags of Canada, France, Germany, Japan, the U.K. and the U.S. (Velocci, 1996).

Fourth, offshore suppliers can be a competitive alternative to U.S. suppliers. As they gain experience, offshore suppliers are becoming increasingly competent and cost effective in meeting the exacting requirements of aerospace subcontracting. Moreover, the mid-1990s global slump in aerospace industry sales left excess capacity worldwide, creating additional opportunities for attractive subcontracting prices.

EXAMPLE STRATEGIC SOURCES: PACIFIC RIM SUPPLIES

Under pressure from its largest customer to expand its use of foreign suppliers, in 1986 GAMC contacted five leading Japanese aerospace groups, seeking bids for work on a structural subassembly GAMC had been doing in-house for its major commercial customer. These assemblies account for approximately 25 percent of the overall weight and 15–20 percent of the cost of GAMC's delivered structures. GAMC selected one Japanese supplier and the first delivery was set for 1987. The next year GAMC also began using a second Pacific Rim source for the same assemblies. When an "act of God" disrupted the parts flow from one of these international suppliers, GAMC's dual-source arrangements enabled it to continue its final assembly production without disruption.

There were up-front costs to GAMC for off-loading this assembly to new suppliers. Because GAMC remained responsible for the final complete assembly sent to its customer, and because they had never done business with either Pacific Rim supplier before, GAMC worked with them to ensure that the subassembly continued to meet their customer's expectations. No estimates for how much GAMC invested in these efforts apparently were made, but GAMC had to transfer its technical know-how on manufacturing procedures needed to meet the customer's requirements. GAMC engineers trained the suppliers on, for example, rivet installation requirements, methods of drilling holes, shank allowances, tolerances, preferred methods for applying sealant, and inspection and quality control procedures. Tips on practical details were passed on, such as air pressures for pneumatic drills to ensure close tolerances and techniques for avoiding air bubbles while mixing sealant. GAMC supplier management personnel also worked to enable the strategic source suppliers to take over responsibility for managing and overseeing the lower-tier parts and materials suppliers that had previously been shipping directly to GAMC for the subassembly.

The relationship with the Japanese subcontractor has expanded since that time. In the mid-1990s, GAMC began off-loading a similar assembly it had also been producing in-house for many years for another commercial program for the same customer, this one its largest single program.

Most recently, GAMC entered into life-of-product design and manufacturing agreements for the COM-Z program with a second commercial customer and with the same Japanese strategic source. This time, from the beginning the strategic source has been responsible for selecting and managing its lower-tier parts and materials suppliers. Leveraging GAMC's established relationship with this Japanese subcontractor was attractive to the second commercial customer because Japan was a potentially large market for the COM-Z, its new transpacific aircraft. Indeed, the international risk and revenue sharing team for the COM-Z includes suppliers throughout the United States, Europe, and several Pacific Rim countries.

The first foreign strategic subcontracting in 1986 actually predated GAMC's formal "strategic sourcing" program. The goals and approach of the formal program, however, are similar. One difference now, however, is that instead of seeking bids, GAMC develops a "target price" with potential foreign suppliers. The price is based on input costs in the target country, combined with GAMC's estimates of its own labor and equipment hours for manufacturing the subassembly under consideration.

Generally, GAMC hopes to see a lower unit cost than in-house. However, foreign sources have also been selected for strategic reasons when there was little difference in costs. For example, in 1994 GAMC began moving two of its large commercial structural subassemblies (again along with responsibility for oversight of lower-tier suppliers) to a European supplier even though their costs were similar. Why? One of their customer's largest commercial markets outside the United States was in the supplier's home country.

Once a subcontractor is selected, GAMC forms what it calls a "buy-to-package team." In the spirit of IPPD, the team, headed by the procurement organization, has representatives from the various functional organizations as well as from the subcontractor. The team provides technical and managerial assistance to the subcontractor to ensure a successful transition of the work. After the lean model of collaborative supplier relations, the goal is to help the supplier improve while at the same time passing down all responsibility for the subassembly. This includes quality control and managing lower-tier suppliers of parts and materials (even those that GAMC previously may have used on the same part). One team participant described how GAMC works with the subcontractor:

> We do have them participate on the team. We ask them to send an engineering crew on the front end to evaluate it completely, the tooling, the engineering data, all that... We bring them in, we give them everything and anything they want to look at: engineering data, we give them access to our shop floor. We have some controls. It does get a little anxious [on our shop floor] when people know that there is a possibility that they'll lose that assembly and then they'll lose their job. So we need to be attuned to what is going on on the shop floor and have to be very sensitive to it.
>
> Our objective is to reduce [the subcontractor's] first unit [costs]. We want the hours to be down, so that when they start learning they recover a lot quicker. Everything we try to do is let them know how we did it, let them know where we think they can improve, and help them along in that matter.
>
> We do a lot of program reviews, make sure they are following guidelines, meeting their milestones, and so on. We do a lot of on-site support. Not only short visits, but we may have resident people at the subcontractor's facility. We do things like facilities evaluation. If they are taking on a new job and they are somewhat puzzled about the floor space, we help them out and tell them what the floor space it is occupying in our facility.
>
> It's all a process of functional organization support, and ultimately of subcontract management. Once the subcontract is in place the procurement organization manages that subcontract to ensure that we get quality and delivery on time, on schedule.

The assistance also sometimes includes training in program management techniques such as total quality management, statistical process control, or integrated product teaming. In addition, because the sources have been selected to support the company's customers, the customers may also be involved in developing the supplier:

> We work very closely with our customers in the development of that subcontractor, because they in turn may be trying to develop that same subcontractor. So we work together to be sure we don't duplicate things. One example is where [our customer] provided composite training to a small company. They left [the customer's] facility, came here and received program management training from us. So, we work very closely with our customers in that respect.

In short, considerable effort and cost goes into supplier development in support of GAMC's international strategic sourcing. All of this effort is geared toward commercial programs. Although there may be future connections to military work, the more general dual-use approach of the company has not yet applied to international strategic sourcing. Indeed, the government's preference is clearly for a "buy American" strategy, driven by the politics of defense spending and by security concerns about the industrial base and guaranteed access. Although offset agreements for foreign content are common when U.S. military products are sold overseas[3], this has not affected GAMC because its primary military programs are not sold internationally. The closest analogy on the defense side of the business would be suppliers chosen because of their location in particular states or congressional districts. But GAMC has no "political sourcing" supplier development program, and managers were, understandably, reluctant to discuss how important such issues are in supplier selection or any specific examples.

NEW ROLES AND RESPONSIBILITIES IN SUPPLIER CHAIN MANAGEMENT

The movement at GAMC toward strategic sourcing, and more generally toward the IPPD philosophy in working with suppliers, has changed the nature of supplier management and materiel and procurement operations at the company. The teaming approach is similar to that pioneered by Toyota in the automobile industry. Indeed, GAMC managers are explicit in saying that much of their organizational philosophy comes from their understanding of research on the Toyota system. As seen in each of the strategic sourcing examples above, GAMC is applying the approach not only internally but also externally in its interactions with its customers and suppliers, both domestic and international.

The changes stem mainly from GAMC's response to the globalization of markets and industrywide contraction since 1991. Increased competition for new contracts led to a re-evaluation of GAMC's market niche and core manufacturing competencies. Corporate strategy shifted from seeing GAMC as a subcontractor to envisioning GAMC as teaming with final assemblers, designing, manufacturing, and integrating complete substructures, and at the same time teaming with lower-tier suppliers. As one manager put it:

> What we hope to do is, gradually, build the large structures, and have support partners who are feeding us good sized subassembly, where we do the integration here, and possibly go one step further and maybe put some systems in the product, and then ship it to our customer. So we want to take the next step up from being a major subcontractor to be more of a support partner [for the final assemblers].

GAMC's ongoing internal evaluation of its core competencies guides decisions about which products and processes to subcontract out and which to retain in house. GAMC has developed a systematic process for determining what it's strategically important core processes are. Cross-functional teams develop "activity maps" to benchmark GAMC's capabilities in each area relative to competitors and potential suppliers. The goal is to identify processes and products for

[3]For example, the F-16 is co-produced in, among other countries, Korea by a group led by Samsung, and 36 British companies are building parts of the C-130J, in large part because it is in competition with the planned European F.L.A. airlifter for purchases by the British military (*The Weekly of Business Aviation,* February 27, 1995).

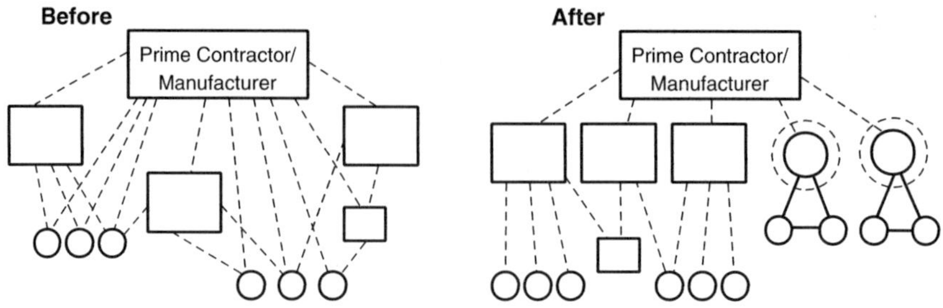

FIGURE 9 Tiering and teaming in supplier management. Source: Nishiguchi (1994).

which GAMC no longer has the volumes to justify maintaining in-house capabilities, or where GAMC is not cost competitive.

At the same time GAMC has moved to reduce the number of suppliers it has to manage and, following a "tiering" strategy, it has begun gradually pushing down to suppliers the responsibility for managing lower-tier suppliers of components and raw materials. We have seen examples already in the context of strategic sourcing. When the process is complete, suppliers inherit responsibility for managing all the component and raw materials in the offloaded subassemblies.

The strategy also means developing longer-term collaborative "teaming" relations with suppliers which GAMC believes are capable of taking on larger responsibilities, and helping the company be both cost competitive and strategically placed for future contracts. A subcontract manager described the concept:

> When we pick a supplier, we don't want to pick a supplier and just give him one component, and that's all he's going to provide us for the balance of our relationship. We want somebody that we can develop, work with, so that when we pursue new programs we know they have the capability. . . . So that if we pursue another contract with our customers, and it involves a similar structure, then we can invite them to participate with us. . . . We don't have to start from scratch.

Figure 9 is a representation of this combination of "tiering" and "teaming," toward which GAMC is moving with both its customers and suppliers. As indicated above, GAMC's new COM-Z program is modeled on this tiering and teaming approach. The approach is also central to Japanese and (some) European automobile subcontracting networks (Nishiguchi, 1994).

But the transition in the U.S. aircraft industry will take many years, in light of the long product life cycles. For example, GAMC has been supplying structures on one program since the 1960s. As a result, GAMC supplier relations look more like the "before" diagram than the "after." The process has really just begun.

The activities described above illustrate a trend among airframers to move certain of their assembly and integration functions downstream to their major subcontractors. Undertaken by the airframers to achieve cost saving, streamlining, and marketing goals, this trend gives subcontractors like GAMC positive benefits in the form of revenue sharing and hands-on experience in design and development teaming. At the same time, the trend imposes more responsibilities and risks as well. As we've seen, one effect has been to increase the subcontractors' roles and responsibilities in supplier chain management. GAMC is expected, in the tiering structure, to fully manage and coordinate the next-tier suppliers. In response, GAMC has taken on new roles, adding a supplier certification and rating system, with a preferred supplier program. It added a "supplier process management" group for providing technical assistance to suppliers. It added the strategic sourcing group discussed above for developing international sources as offsets in support of customer marketing efforts. Moreover, because GAMC integrated its supplier system across all programs, these supplier-related efforts are complicated by the fact that each customer has its own set of requirements and preferences, its own supplier rating and certification systems, and its own set of approved and preferred suppliers. As a supplier also responsible for an extensive supplier network of its own, GAMC's materiel and supplier management operations took on increased and changing responsibilities from two directions.

The result was that the whole set of "buying" activities at GAMC moved much further toward oversight and management roles and away from the procurement tradition of price negotiation and contracting. Negotiation and contracting activities continued, of course, but pricing was complicated by "best value" supplier selection and "target pricing." More important, in the IPPD approach, materiel and supplier management personnel take on additional responsibilities such as liaison and coordination activities between suppliers and internal functional groups working with suppliers on process and product improvement activities. GAMC materiel and supplier management personnel also oversaw and coordinated teaching, monitoring, and evaluating suppliers, and as the quote above suggests, even managed in-house morale and union tensions as the company moved to outsource internationally.

These added responsibilities also require an expanded set of skills in the materiel functions. For example, the strategic sourcing initiative meant finding people with international experience and an understanding of managing the risks of international markets, as well as of business practices and cultures in the Far East and elsewhere. Several people at GAMC have "China" in their titles. Negotiation and supplier selection by buyers no longer is simply a matter of finding the best price. Rather it often requires sound technical judgment and an understanding of how the supplier will fit within GAMC's internal functional activities. Different organizational and people skills are needed, too, because GAMC buyers participate in cross-functional teams. This means communicating with engineers and designers and human resource people, and sharing responsibility for the team's performance.

The changes have also increased the need for expertise and resources in in-

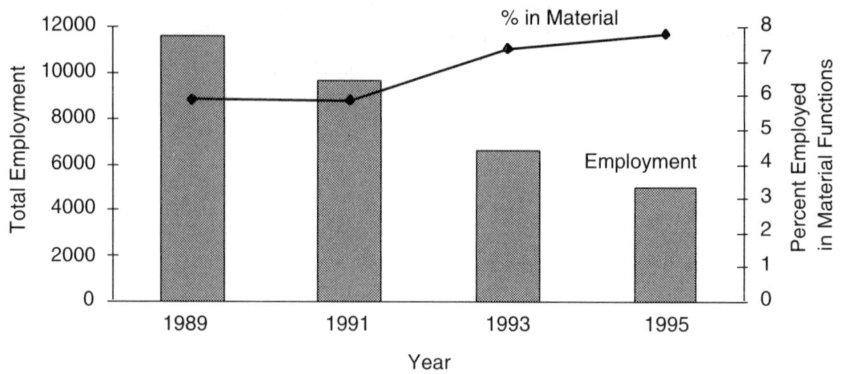

FIGURE 10 Employment at GAMC and material functions as percent of total.

formation management. GAMC's supplier rating and certification systems must efficiently gather, process, and organize performance metrics such as on-time deliveries, defect rates, statistical process control data, and supplier financial information and make it available in a timely way to the appropriate decision makers. Outsourcing "make–buy" decisions cannot be made without good internal cost accounting information. In the cooperative supplier relations model, advanced scheduling information is gathered from GAMC's functional organizations and program teams and then processed and made available to suppliers for planning purposes.

As a result, personnel engaged in materiel functions at GAMC as a fraction of overall employment increased between 1989 and 1995. As Figure 10 shows, although the approximately 400 people performing material functions at GAMC before consolidation with its new owners (including subcontract management, supplier development and technical support, inventory control, purchasing, administration, procurement quality, receiving, warehousing, shipping, and off-site personnel) was much reduced from 685 in 1989, employment in these functions fell less precipitously than overall employment.

RISKS FROM INTERNATIONAL COMPETITION

The combination of the increased importance of foreign commercial markets compared with U.S. commercial and defense markets, and the implementation of collaborative supplier relationships, poses an additional significant dilemma for the major structural subcontractors like GAMC, and perhaps for all suppliers in the U.S. aircraft industry. If GAMC is representative, the combination of market globalization and lean supplier development strategies implies that the major subcontractors outsource to the strongest foreign firms, particularly in the Pacific

Rim, and work with those foreign firms to improve them. Are the major U.S. subcontractors like GAMC, through their technology sharing and supplier development practices, helping to create their own overseas competitors?

For example, one of GAMC's customer's newest commercial programs includes one of GAMC's original Pacific Rim strategic sources but not GAMC, despite the company's long history with that customer. It is obviously impossible to determine the extent to which GAMC's supplier development efforts contributed to that customer's sourcing decisions. GAMC's absence is, in part, explained by differences in technology. Their customer developed proprietary manufacturing processes for use on the new program. Although GAMC possessed similar capabilities, their customer elected to use this new technology rather than subcontract that portion of work to other suppliers. Whatever the reason, GAMC's Pacific Rim strategic source has work on the customer's new program and GAMC does not. Ironically, GAMC was instrumental in introducing the Pacific Rim supplier to its commercial customer in the first place.

Similarly, another of GAMC's strategic source Pacific Rim suppliers has a widely reported and very clear strategy across all its businesses, from chemicals to electronics to automobiles to aircraft: "acquire technology abroad, then go independent." It acquired robotics technology by jointly producing with other equipment manufacturers and now makes its own. It licenses automobile technology, but plans to go alone by 2005. It bought leading electro-optical technology firms in Europe and the Pacific Rim. It is now aggressively pursuing teaming with international partners in order to use its experience as a second- and third-tier aerospace supplier to eventually compete directly with the major airframers, the top design and final assembly tier of the aircraft industry.

GAMC managers report that, in their efforts to find foreign sources, few suppliers they approached wanted to do detail fabrication of parts. Rather, the foreign firms wanted to integrate and produce more complex subassemblies. GAMC's experience is consistent with the findings of a U.S. General Accounting Office (GAO) report in 1994 that China, Japan, Indonesia, and Taiwan are all intent on developing their own aerospace industries (U.S. General Accounting Office, 1994). According to the GAO, all are importing product and process technologies and have strong links between military and commercial projects. And their progress has been steady. For example, in the 1960s Korean firms did light repair and aircraft maintenance. In the 1970s they moved up to depot maintenance and contract assembly. In the 1980s they manufactured aircraft parts locally and did increasingly complex assemblies. In the 1990s they have self-developed a small jet aircraft (50 seat) and have moved to do co-production of the F-16. Future plans call for developing a 100 seat jet aircraft.

CONCLUSIONS

Case studies, by nature, do not offer statistically useful hypothesis tests and

TABLE 4 Performance Improvements Reported by GAMC

1. Overall cycle time to customer delivery:
   - 17 percent reduction on commercial programs
   - 14 percent reduction on a military program
2. Product cost:
   - cost per pound improvement on all programs
3. Rework, repair, and fabrication costs:
   - 40 percent reduction per standard hour
4. First time quality:
   - 89 percent of first article inspections on new program passed
   - second upper skin structure on new program was defect free
   - first fully integrated structure mated in two hours versus three days original job allocation
5. Customers' perception:
   - award for excellence
   - supplier of the year
   - preferred supplier (top two percent)
   - outstanding quality recognition on a military program

cannot control for any other possible explanatory events. Nevertheless, with its emphasis on the whole lean paradigm, GAMC does provide the opportunity to begin to explore the hypothesis that the methods derived from studying the automobile industry will also pay off in the aircraft industry. Because the central focus of this case has been limited to supplier management and international sourcing, it has not explored the broader lean paradigm at GAMC. In particular, a closer and independent investigation of manufacturing practices and costs would be needed to confirm the performance metrics listed in Table 4, which shows GAMC's self-reported improvements on its major programs. The company, at least, believes it has seen significant results from the whole package of changes undertaken in pursuit of lean manufacturing practices.

More specific to supplier management, the lean paradigm derived from research on the automobile industry suggests several practices. As we have seen, GAMC embraced and implemented many of them, including increased outsourcing of non-core manufacturing activities; teaming with suppliers and customers in design and development; working with suppliers to improve their technical and managerial capabilities; enhancing information flows and technical exchange with suppliers to improve joint problem solving; and reducing the number of suppliers through working only with the best ones.

These trends increased the roles, responsibilities and costs (as a fraction of overhead) of GAMC's materiel functions. They also began to strain the integrated dual-use approach to supplier management. However, the dual-use approach will likely remain a viable alternative to separate program or divisional supplier management. The advantages, outlined above, of the dual-use approach appeared to outweigh the problems. GAMC continued to use a single MilQ-approved procurement system for both military and commercial programs with very few deviations for commercial reasons. Moreover, duplication of materiel functions is not likely to be attractive in the future, either, in an era of intense commercial competition and significantly reduced levels defense spending. So too, the Pentagon has increased its preference for "commercial practice" both in manufacturing and contracting.

However, if GAMC is representative of the broader supplier base, the upper supplier tiers of the U.S. aircraft industry are getting squeezed between, on one hand, the push toward lean practices, which increase supplier responsibilities and risk, and, on the other hand, the global sourcing and offset initiatives of top-tier companies. GAMC has no real choice but to do everything it can to meet its customers' needs. GAMC's customers may have no real choice but to do everything they can to increase their foreign sales. The commercial side of GAMC's customer base may indeed do very well by increasing the pressure for foreign sources. Yet even at the top tier there is risk together with the potential returns in combining foreign offsets with the strongly developmental practices suggested by lean mantras of collaborative supplier management. It is clear that with the experience gained from ever-increasing responsibility in the aircraft supply chain, Pacific Rim competition is on the horizon not only for GAMC-like first supplier tier assembly work, but also for entire 100-seat transport aircraft. The Pentagon, too, must be concerned today about the effects of this offset–lean squeeze on the health of the U.S. aircraft industrial base. The squeeze may end up suffocating the vital middle tiers.

ACKNOWLEDGMENTS

Support for this research was provided by the Lean Aircraft Initiative at MIT. Analysis and conclusions expressed herein are solely the responsibility of the author and not of the GAMC, its parent company, or the Lean Aircraft Initiative, the other sponsoring organizations, or MIT. The quotes within represent the opinions of personnel interviewed by the author, but not necessarily of these organizations. These disclaimers aside, I must express my sincere gratitude for the friendliness and cooperation, not to mention their countless valuable hours, with which GAMC personnel have been willing to talk with me about these issues. I have found both a warm reception and knowledgeable people throughout the enterprise, vice presidents to shop floor operators. The case study would not have been possible without that help. Thanks also to Maryellen R. Kelley for sage advice during preparation of the case, and to members of the LAI Supplier Systems and Relationships team for comments on an earlier summary of initial findings.

REFERENCES

AIA (Aerospace Industries Association). 1994 and 1995. *Year-End Review and Forecast*. Washington: AIA.

Boeing Company. 1990. Press release. December 12. Boeing Company, Seattle, Washington.

Helper, S. 1991. How much has really changed between U.S. automakers and their suppliers? *Sloan Management Review* (Summer):15–27.

Kelley, M.R., and T.A. Watkins. 1995. In from the cold: prospects for conversion of the defense industrial base. *Science*, Vol. 268:5210: 525-532.

Lyons, T.F., A.R. Krachenberg, and J.W. Hands, Jr. 1990. Mixed motive marriages: what's next for buyer-supplier relations? *Sloan Management Review* (Spring):29–36.

McMillian, J. 1990. Managing suppliers: incentive systems in Japanese and U.S. industry. *California Management Review* 32 (Summer).

Nishiguchi, T. 1994. *Strategic Industrial Sourcing: The Japanese Advantage*. New York: Oxford University Press.

U.S. Department of Commerce. 1994. *1994 National Trade Data Bank, Market Reports*. March 17. Washington, D.C.: U.S. Department of Commerce.

U.S. General Accounting Office. 1994. *Asian Aeronautics: Technology Acquisition Drives Industry Development*. Washington, D.C.: U.S. General Accounting Office.

Velocci, A.L., Jr. 1996. "Global express vs. G5: the contest heats up. *Aviation Week and Space Technology*, September 2.

Womack, J.P., and D.T. Jones. 1994. From lean production to the lean enterprise. *Harvard Business Review* (March–April):93–103.

Womack, J.P., D.T. Jones, and D. Roos. 1990. *The Machine that Changed the World*. New York: MacMillan.

The Role of the United States Government in Setting Offset Policy

Owen E. Herrnstadt[1]
International Association of Machinists and Aerospace Workers

INTRODUCTION

Many countries force U.S. aerospace companies to transfer high-skilled jobs and valuable technology to them in return for the purchase of U.S. aerospace products. In order to gain market access in countries with "direct" offset requirements, U.S. based contractors compensate these countries in some form "directly related to the system being exported".[2] For example, in return for purchasing jet fighters or commercial airplanes, U.S. contractors agree to produce part of the jet fighter or the commercial airplane in the purchaser's country. In addition to direct offset requirements, countries are increasingly requiring U.S. contractors to satisfy "indirect" offset requirements that include compensation in forms that are "unrelated to the exported item".[3] Under this scheme, for example, instead of forcing a transfer of defense production, U.S. contractors would rely on the commercial aerospace industry to satisfy military offsets. In some cases, non-aerospace industry products, are also relied on to satisfy offset arrangements.

Although offset arrangements are traditionally linked with a country's trade practices, similar arrangements are becoming more common through voluntary

---

[1]Director for the International Affairs Department, International Association of Machinists and Aerospace Workers. This paper is based on a presentation made at a National Research Council Workshop on offsets held on June 9, 1997, in Washington, D.C. The views expressed herein are those of the author and do not necessarily reflect the views of the IAM. It should be noted that while this paper focuses on offsets in the aerospace industry, offsets in other industries require similar attention.

[2]Trade Promotion Coordinating Committee, National Export Strategy: *Toward the Next American Century: A U.S. Strategic Response to Foreign Competitive Practices*, U.S. Government Printing Office, Washington, D.C. 1996, p. 155.

[3]*Ibid.*

agreements reached between private parties. These "offset-like" arrangements may include a direct or indirect type of offset arrangement. While for policy purposes, a distinction as to what kind of offset is involved may be important, for workers who face their negative effects, distinctions are of little relevance. For purposes of discussion in this paper, the term "offsets" is used broadly, to include both direct and indirect offsets and offset-like arrangements "voluntarily" agreed to by private parties.

While these sophisticated offset policies and marketing schemes are increasingly utilized by other nations to promote the development of foreign aerospace industries, the United States demonstrates little interest in developing a comprehensive policy of its own. But the U.S. government can no longer afford to leave the world of offsets to the actions of other nations and private parties. The stakes are too high. Offset arrangements in the defense and commercial aerospace industry result in the loss of U.S. jobs and technology to other countries. In some cases, they can pose a threat to our national security.

Consequently, government must play a strong role in developing policies that address the rapid acceleration of offsets in the aerospace industry and their negative effects on U.S. aerospace workers. In addition to the projected loss of thousands of jobs, over time the effects of these arrangements could result in the decline of the U.S. aerospace industry, one of our greatest remaining export industries. This paper examines why the federal government must take a leadership position in setting offset policy in the U.S. aerospace industry by reviewing the health of the aerospace industry from the view of the aerospace worker; the increasing threat offsets pose to aerospace workers and the national interest; the serious lack of current *and* accessible information on offsets, the need for coordination of offset policy within government, and the need for coordination of offset policy between the numerous private parties that are involved either directly or indirectly with offsets.

THE PROBLEM

The U.S. Aerospace Industry Worker Faces a Gloomy Future

The impact that offsets have on the U.S. work force receives little attention from public policy makers. Periodically various labor statistics are quoted in articles about offsets. However, few people have focused on the effects that offsets are having, and will increasingly have, on the lives of real workers, their families, and the communities where they reside. The condition of these workers is especially important for a number of reasons.

First, U.S. aerospace workers are in large part responsible for building the U.S. aerospace and defense industries and for making them the leaders that they are in the world today. U.S. aerospace workers are loyal and proud of the companies they work for and the communities that they live in. They share common desires to have secure employment, earn reasonable wages, and receive decent benefits. They also want to work in a safe and healthy environment and spend quality time with their families. We owe our allegiance to them, just as they have given their allegiance to us.

In addition, preserving and expanding decent jobs in the aerospace industry makes good economic sense. These jobs are generally high-skilled, high-wage jobs. Consequently, the Clinton Administration has recognized their importance in today's economy. In his opening remarks to a National Research Council workshop on offsets, Gene Sperling, Director, White House National Economic Council, "stressed that the goal of the Administration is to develop the best policy to create high-wage jobs for American workers . . . Job retention and job growth in the aerospace industry is important to achieving the overall goal of a more secure and higher-paid workforce."[4] Although the Administration's goal is highly laudable, much work needs to be done in order to ensure that it is achieved.

Employment Prospects and Income Effects

Employment prospects for U.S. aerospace workers are troubling. One estimate concludes that between 1990 and 1994, the U.S. lost roughly 500,000 jobs in the U.S. aerospace industry and roughly 1 million other jobs which are "dependent" on the U.S. aerospace industry.[5] This represents a staggering decline of almost 40% of U.S. aerospace employment for the five year period. For aerospace workers who have lost their jobs and for those who risk losing their jobs in the coming years, the decline in employment is painful.

The International Association of Machinists and Aerospace Workers, (IAM) the labor organization which represents the largest number of workers in the aerospace industry, conducted a 1996 survey of its members who had lost their jobs in the aerospace industry.[6] The survey involved displaced workers from two large aerospace and defense companies, Lockheed-Martin in Marietta, Georgia and the Boeing Company in Seattle, Washington. The results indicated that at the time of the survey only a small group of aerospace workers who were laid off remained employed in the aerospace industry (16.8 percent).[7] The average displaced aerospace worker reported earning nearly $3 less per hour than they earned in their

[4]Summary of Comments of Gene Sperling, Director, White House National Economic Council, *Policy Issues in Aerospace Offsets: Report of a Workshop* (hereinafter referred to as "*Workshop*"), National Academy Press, Washington, D.C., 1997, p. 1.

[5]Randy Barber and Robert E. Scott, *Jobs on the Wing: Trading Away the Future of the U.S. Aerospace Industry*, Economics Policy Institute, Washington, D.C. 1995, p. 1.

[6]IAM Strategic Resources Department, *IAM Survey of Displaced Aerospace Workers*, November 1996.

[7]*Ibid.*, p. 2.

previous aerospace job.[8] Nearly half of the respondents to the survey indicated that they were earning less than 75 percent of their previously hourly rate.[9]

Not surprisingly, this dramatic decline in income has had a substantial impact on workers' families. In some cases, it has significantly changed the family structure. An increasing proportion of the spouses of respondents were forced to find employment. Approximately one-third of spouses who were not working prior to the layoff were working at the time of the survey.[10]

In addition to their drastic reductions in income, displaced workers also lost valuable benefits. For example, roughly 75 percent of respondents lost health insurance coverage when they were laid off.[11] The majority of respondents indicated that when they finally obtained new employment, their health care coverage was worse, or much worse than the coverage they previously had.[12] Even more troubling was the finding that laid-off workers suffered from an increasing number of physical ailments. A growing number suffered from increases in high blood pressure, heart problems, digestive tract problems, and sleep disorders.[13]

The trauma of suffering a loss of job, particularly in a high-skilled, high-wage industry like aerospace was poignantly conveyed by the words of IAM members who were laid off. One respondent in the IAM survey reported, "[A]fter being laid off, my self-worth has gone to zero. Our financial outlook is bleak. It's very hard to make ends meet, even with two working."[14] Another member explained, "The thing that was so bad, was [losing] the hope of having the chance of gaining anything for old age . . . ."[15]

Indeed, opportunities for decent employment in the U.S. aerospace industry are gloomy. Researchers predict that approximately 250,000 jobs are in jeopardy in the aerospace and related industries by the year 2000, and almost 500,000 jobs at risk by 2013.[16] "Direct jobs lost in 2013 would represent 25.6 percent of the total jobs in aircraft production in 1995."[17]

The Impact of the Asian Crisis

That some aerospace companies are currently hiring high-skilled workers does not lessen the impact that offsets are having and will increasingly have on

[8]*Ibid.*
[9]*Ibid.*
[10]*Ibid.*
[11]*Ibid.*
[12]*Ibid.*
[13]*Ibid.*
[14]*Ibid.*, p. 3.
[15]*Ibid.*
[16]Barber and Scott, *Jobs on the Wing*, p. 2.
[17]Scott, "The Effects of Offsets, Outsourcing, and Foreign Competition on Output and Employment in the U.S. Aerospace Industry," presented to National Research Council Symposium on *Trends and Challenges in Aerospace Offsets*, January 14, 1998.

the U.S. aerospace worker. The recent hiring frenzy in the commercial aerospace industry is taking place in the midst of a booming economy. What will happen when the economy takes a downturn? Moreover, the number of aerospace workers that have recently been hired does not even begin to make up for the massive job losses that have occurred over the last several years. Nor will these recent hirings be able to minimize the negative impact that offsets will have on the U.S. workforce in the future. In addition, claims that there are not enough aerospace workers to fill current job demands may be, in part, due to decisions by former aerospace workers laid off prior to the boom to seek employment in more stable industries. Lastly, what will happen in the aerospace industry if the ripple effect from the Asian financial crisis turn into a Tsunami? Sub-tier producers and their employees in the U.S. may especially feel the brunt, as prime contractors face cancellations and manufacturing costs drop even lower in Asian countries, making bids for remaining work even more competitive.

The increasing reliance on offsets is one factor that is contributing to the gloomy picture of aerospace employment in this country. While no one would argue that the huge layoffs that have occurred in the U.S. aerospace industry were caused solely by the practice of using offsets, it is irresponsible to ignore the serious effects that offsets can have on such an essential industry—an industry that is key to the economic health and prosperity of our country. Consequently, every possible cause for the sharp decline in employment that has occurred or that may occur in the future must be explored.

Offset Arrangements are Becoming a Standard Way To Do Business

The use of offsets as a marketing tool is increasing in both defense and commercial aerospace industries. In fact, since the 1980's, indirect offsets have grown even faster than direct offsets.[18] As offsets take new forms, their effects are blanketing the commercial aerospace industry: "Current information leads to the conclusion that indirect offsets are increasingly the norm."[19] In the defense industry, the use of offsets has undergone "a substantial increase in new obligations over previous years, both in value and as a percentage of export contracts."[20] Although some people contend that offsets open certain markets, their effects can be negatively felt by other areas of the economy. Indirect offsets may have

---

[18]Department of Commerce, *Offsets in Defense Trade: A Study Conducted Under Section 309 of the Defense Production Act of 1950, as amended,* Bureau of Export Administration, Washington, D.C., 1996, p. 71.

[19]Summary of Comments of William Reisch, *Workshop,* p. 20.

[20]Department of Commerce, *Offsets in Defense Trade: A Study Conducted Under Section 309 of the Defense Production Act of 1950, as amended,* Bureau of Export Administration, Washington, D.C., 1997, p. i.

"unfavorable consequences for subcontractors and increased risks for a wide range of companies throughout the U.S. economy."[21]

The U.S. Department of Commerce, Bureau of Export Administration Defense Diversification Needs Assessment Survey program covering the period 1993-1994, found that 83% of subcontractors who responded to the survey that they were "positively or negatively impacted by offsets" had been "harmed" rather than helped by offsets.[22] Furthermore, one-third of the 1,100 transactions examined involved " partial or full production" of the items sold in the country which purchased them.[23] " In many cases, this has led to the creation of redundant or excess defense manufacturing facilities."[24] In other words, these offsets have resulted in over-production capability in the defense industry. To the extent that over-production in the defense industry negatively affects production in the commercial aerospace industry, sub-tier producers that employ thousands of aerospace workers are also affected.

New Entrants Pose Risks to the U.S. Supplier Base

Dr. Kirk Bozdogan of the Massachusetts Institute of Technology also sees the danger of the increased use of offsets. He explained during an NRC workshop that the use of offsets will increase and warns that they "will pose serious risks for the U.S. supplier base."[25]

Indeed, he reports that the U.S. supplier base has decreased drastically (by 50%) in the aerospace industry over the period 1991–1995.[26] He also concludes that foreign programs in offset requirements will lead to too much production capability and warns of increasing costs and competitive pressures on U.S. companies that provide components and other items that support the major aerospace companies.[27]

These findings are consistent with the observations of those who report that numerous countries around the world have well-developed strategies to build their own aerospace industries through offsets. "A number of these competitors have the goal of developing a full-service commercial aerospace industry. China is assembling entire Western designed jetliners . . . . Japan is mounting a systematic effort to become a first-tier aerospace manufacturing power . . . . Even only

[21] *Offsets in Defense Trade*, 1996, p. 71.

[22] *Ibid*, p. 63; Cited in Trade Promotion Coordinating Committee, *National Export Strategy*, 1996, pp. 162–163.

[23] *National Export Strategy*, 1996, p. 163.

[24] *Ibid.*

[25] Summary of Comments of Dr. Kirk Bozdogan, *Workshop*, pp. 27–28.

[26] *Ibid.*, p. 28.

[27] *Ibid.*

recently industrializing countries . . . have joined over thirty other participants in the global contest for a share of aerospace."[28]

In their 1995 study, *Jobs on the Wings*, Barber and Scott document the strength of the growing Chinese aerospace industry and its close ties to U.S.-based aerospace manufacturers:

"China is already working closely with McDonnell Douglas assembling Western-designed commercial aircraft as part of a coproduction arrangement to manufacture 40 MD-82s and recently finalizing an agreement for the production of 20 MD-90 'Trunkliners' in China, with dramatically increased Chinese content (reportedly 85% by the end of the production run)."[29]

Aerospace industries are also burgeoning in Japan, South Korea, Indonesia, and Taiwan.

Among other activities, the South Koreans have attempted the development of an "Asian Airbus", and have had a number of production contracts with U.S.-based aerospace companies.[30] Taiwan has also been advancing toward the establishment of a viable aerospace industry with advantageous offset requirements that insist "that U.S. companies 'promise to allow Taiwan to build aircraft and engine parts, acquire U.S. technology, and receive training and other support for its developing aeronautics industry'."[31] Indonesia's entry into the world's budding aerospace industries is also growing. Its state-owned aerospace company already "produces numerous military and commercial aircraft under licensed production agreements".[32] It also "makes significant parts for all three major aircraft manufacturers."[33]

Insufficient Information About the Effects of Offsets

In general, there is a serious lack of information about offsets and their effects on workers. While the government has limited knowledge of military offsets, it has little knowledge about the nature, extent, and impact of offsets in the commercial aerospace industry. Unfortunately, under the current situation, this type of information is next to impossible to obtain. First, the corporate culture which relies on confidentiality to maintain competitive advantage makes it difficult for workers to obtain information about offsets. If workers were aware of the impact that specific offsets were having or were going to have on them, they could work with the company in an attempt to either avoid or minimize their

[28]*Conflict and Cooperation in National Competition for High-Technology Industry*, National Academy Press, Washington, D.C., 1996, p. 88.

[29]Barber and Scott, *Jobs on the Wing*, p. 62.

[30]*Ibid.*, p. 66.

[31]*Ibid.*, p. 67.

[32]*Ibid.*

[33]*Ibid.*

negative effects. However, if workers are not made aware of the offsets, let alone their effects on the workforce, they cannot offer their expertise in alleviating the harmful effects that offset arrangements can have on their jobs.

One of the consequences of a company's failure to inform its workers fully about their offset arrangements is that it fosters distrust and other forms of ill-will. If workers are not told the truth about corporate marketing schemes they may assume that the company is not protecting worker interests. This is especially true if a company moves production overseas as part of a lucrative transaction that is the result of an offset arrangement. Distrust is also generated by false employer claims that job reductions are a result of business downturns when they are really a consequence of offsets.

An arrangement that was made by one company in the early 1990's illustrates these problems.[34] It had entered into a co-production deal for F-16's with South Korea. "This deal provided for South Korea to purchase a total of 120 aircraft, of which 72 would be manufactured and assembled in South Korea, 36 would be assembled from kits in South Korea, and only 12 would actually be made by [the company's] workers in Fort Worth."[35] From the workers point of view this offset arrangement was bad enough, but how the company sought to fulfill its offset arrangement made things worse.

According to the union which represented workers at the company, the company wanted to bring 500 South Koreans into the plant to train them while at the same time, approximately 3,000 union members at the facility were on lay off.[36] The union objected. "The protest put a stop to this scheme . . . we thought. But then we learned that [the company] simply arranged for these . . . workers [from Korea] to be trained at the F-16 plant in Turkey!"[37]

Information about offsets is also difficult to obtain because the fundamental nature of offsets makes information difficult to track. The mere fact that offsets are spread throughout the world taking a multitude of forms makes it extremely difficult to gather complete information. While offsets can have immediate impact on prime contractor production, their effects on sub-tier producers may be harder to trace. When indirect offsets, that affect a multitude of aerospace and non-aerospace companies, are involved it becomes even more difficult to track the offset's effects. This is especially true if companies who are engaged in offsets do not make the effort to monitor the direct and indirect effects of their own arrangements.[38]

---

[34]See, Barber and Scott, *Jobs on the Wing*, p. 37–38.

[35]*Ibid.*.

[36]*Ibid.*

[37]*Ibid.*, p. 38.

[38]See, Summary of Comments of Carol Evans, Workshop, p. 14.

Offsets Can Lead to Conflicts

Critics of a strong government role in setting policy on offsets and similar marketing schemes often contend that government intervention is unnecessary because corporate interests do not conflict with public interests. They fear that unless they are permitted to freely engage in offsets, markets will remain closed, and overseas competitors, who are eager to meet offset demands, will displace U.S. aerospace companies as the leading suppliers of aerospace products throughout the world. They argue that this can only result in harm to the United States aerospace industry and the U.S. economy. They argue that threats of massive job losses and threats to national security because of offset arrangements are unfounded. Job losses, they say, will be made up by increased market share in the world economy. Moreover, they claim that national security will not be threatened because they are careful not to transfer sensitive technology abroad.

In today's global economy, multinational companies know no boundaries.[39] In fact, this is precisely why they *are* multinational companies. Consequently, the loyalty of U.S.-based multinationals to the U.S. may not be as strong as some would have us believe. After all, numerous U.S.-based corporations have moved production facilities to other countries in the search for lower labor costs as well as other perceived "advantages" available in other countries, leaving workers who devoted their careers to them without jobs and the communities which fostered their growth empty.

Market access is essential, especially in high technology trade. Offsets, which enable companies to gain access to other markets, might in some cases be necessary compromises. However, where possible, the U.S. government should eliminate these market restraints. In any event, offsets must be limited when the public interest is jeopardized. There has already been a prior discussion regarding the public's interest in minimizing the social and economic effects from job losses that occur as the result of offset arrangements in the aerospace industry. Two other issues resulting from the aerospace industry's growing reliance on offset arrangements concern the public's essential need for national security and the necessity of balancing the impact of one corporation's offset arrangements against the interests of another corporation.

National Security Impacts

Offsets threaten the national security by fostering proliferation of defense systems abroad and by shrinking the essential sub-tier defense production base at home.[40] The use of offsets increase the capability of developing countries to

[39]See, William Greider, *One World Ready or Not: The Manic Logic of Global Capitalism*, Simon & Schuster, 1997.

[40]Summary of Comments of Carol Evans, Workshop, pp. 14–15.

produce their own weapons systems. Even seemingly minor kinds of aerospace offsets can aid in the development of a weapons system by such things as enhancing "the platforms used for the delivery of chemical or biological weapons."[41] Thus offsets can help to expand the defense capabilities in developing countries thereby creating a greater military threat to the U.S.[42]

In addition, assisting the development of a defense industry in other countries may have the "spiraling" effect of encouraging these countries to seek additional offsets to further supplement their defense production capabilities.[43] "Offsets adversely affect the U.S. supplier base by aiding foreign competitors at the same time that the supplier base is being hit by shrinking defense budgets. Shrinking budgets then lead to a further squeeze on suppliers to give even more offsets."[44]

Examples of national security problems that have been caused, at least in part, from offset arrangements abound. Technology transferred to Brazil through an offset resulted in an improvement of targeting capability of the Iraqi Scud missile system.[45] Under another offset arrangement, McDonnell Douglas sold machine tools to the China National Aero-Technology Import and Export Corporation to be used for production of commercial aircraft.[46] Some of the tools, however, were transferred to the Nanchang Aircraft Company which produces Chinese military equipment.[47]

Domestic Impacts

Offsets also harm other U.S. domestic companies that operate in the industry participating in the transaction. For example, prime contractors trying to expand their access to international markets are unlikely to be concerned by the effects on domestic sub-tier producers whose sales might be substituted for foreign goods as part of offset arrangements. Transferring production of one piece of a defense system to a producer in another country may be inconsequential to a prime contractor, but to a subcontractor, who is able to concentrate on only a few programs, it could be fatal.[48] The following comments which were received when the Defense Diversification Needs Assessment Survey was conducted illustrate how the offsets which "benefited" one aerospace company affected another aerospace or aerospace-related company:[49]

---

[41]*Ibid.*

[42]*Ibid.*

[43]*Ibid.*

[44]*Ibid.*

[45]*Ibid.*

[46]U.S. General Accounting Office Report to Congressional Requesters, *Export Controls: Sensitive Machine Tool Exports to China*, November 1996.

[47]*Ibid.*

[48]See Summary of Comments of Chip Block, *Workshop*, p. 33.

[49]*Offsets in Defense Trade*, 1996, pp. 63–64

• A world-class aerospace and naval forging manufacturer in the Midwest stated that they had "lost significant amounts of work due to prime contractors utilizing foreign sources to satisfy offset requirements."

• A northeastern precision aerospace machine shop reported, "[W]e've lost 20 percent of our business to mandated offset agreements. In the future this will grow substantially. This is our number one problem."

• A manufacturer of rolled rings for aerospace applications stated, "[O]ur company has been significantly affected by [prime engine contractor's] offset agreements to Asia and Europe. I estimate that our company has lost more than 50 percent of our business due to offset agreements."

• A west coast machine shop reported, "[W]e've lost processing work on the jobs that went overseas as a result of aircraft and military hardware sales." Another aerospace machine shop stated, "[T]he aerospace prime contractor we supply] participates in an offset program which seems to have introduced increased competition and possible lost orders to American manufacturers."

• A midwest company that designs and manufactures pumps and valves for aircraft applications reported, "[N]ew competitors created as a result of offsets. Foreign countries now designing indigenous aircraft using this technology."

• A western producer of castings for commercial, aerospace, and defense industries reported, "[N]ew competitors were created or strengthened due to an offset program, hence, we lost the contracts."

The Impact of Indirect Offsets

The growing and innovative use of offsets also creates conflicts between corporations in different industries. In one situation, "Northrop Corp. offered $1.5 million to persuade a U.S. company to buy a $50 million papermaking machine" from a Finnish company.[50] A competitor, based in the U.S. who also makes paper-making machinery, had wanted the sale. "The offer followed a promise by Northrop to the government of Finland to produce American customers for Finnish goods if Finland would purchase F-18 fighter jets from the U.S."[51] Afterwards, a 1994 law was enacted regulating such transactions, by prohibiting "certain types of incentive payments related to offsets".[52]

Last year U.S. Senator Russell Feingold (Wis.), who authored the 1994 law, asked the U.S. Department of Justice to investigate an alleged violation by McDonnell-Douglas.[53] He was specifically concerned that McDonnell-Douglas may have used a tactic similar to the one used by Northrop to satisfy an offset

---

[50]As reported in Aerospace Daily, *Justice Dept. to investigate if McDonnell Douglas broke offset law*, January 3, 1997, p. 15.

[51]*Ibid.*

[52]*Ibid.*

[53]*Ibid.*

arrangement.[54] He feared that a U.S. company lost out on a sale.[55] Senator Feingold summarized the conflict: "It is difficult to be competitive when you are being outbid by foreign competitors assisted by huge defense firms. . ."[56]

It is not just lack of coordination between private entities that are of concern. There also appears to be a serious lack of coordination between public entities. For example, while the Administration seemingly grapples with this issue, the Federation Aviation Administration (FAA) issued regulations concerning "Fees for Providing Production Certification-Related Services Outside the United States." Basically, the FAA's regulation makes it possible for it to sell its services to facilities located outside the United States. The FAA noted in its proposed regulations that some of the "advantages" received by engaging in production of "complex parts, sub-assemblies, or products" outside the United States include:

1. Taking advantage of lower labor costs; and
2. Fulfilling certain aircraft purchasing requirements that require a production approval holder to produce a percentage of the aircraft within the purchasing country.[57]

Somehow the FAA, believes that such a rule would not impose a significant cost impact "on a substantial number of smaller entities." It also dismisses concern that implementation of the proposed rule could have a serious negative impact on U.S. aerospace workers.

The FAA's conclusion flies in the face of studies which conclude the danger of relying on offsets. And while in its final rule, the FAA claims the new rule itself "takes no position on the use of offsets,"[58] the FAA also clearly states that it "recognizes that the indirect effect of this rule may increase the use of facilities and suppliers outside the United States."[59]

A FRAMEWORK FOR PROVIDING A SOLUTION.

Given the negative and growing effects of offsets on U.S. aerospace employment, the lack of accessible information on offsets, the lack of coordination within the U.S. government, conflicts between corporate interests and public interests, a comprehensive national policy on the use of this trade mechanism is needed.

Current efforts are inadequate. While the government, though its Trade Promotion Coordinating Committee (TPCC) Report should be commended for acknowledging that offsets are growing and that there is a woeful lack of informa-

---

[54]*Ibid.*
[55]*Ibid.*
[56]*Ibid.*
[57]U.S. Federal Register, Volume 62, Number 135.
[58]U.S. Federal Register, Vol. 62, Number 207.
[59]*Ibid.*

tion about offsets in the commercial aerospace industry, it is simply not enough to merely note that offsets are increasing and that there is a lack of information about them. The Administration must also acknowledge the serious threats that offsets have on producers, U.S. workers, and the communities in which they reside. And, of course, the Administration must also accept that it has a strong role to play in developing a national policy on offsets.

A stronger role for the Administration means much more than what is called for in the 1997 Trade Promotion Coordinating Committee report. In that report the TPCC describes its meager plan for 1998, which is centered on more ad hoc consultations involving affected groups with the somewhat vague notion of establishing "mechanisms that will both encourage a consensus among the various interests on this issue and provide TPCC agencies with an inventory of information that could be used as a basis for determining the impact of offset requirements and whether any U.S. government action is warranted."[60] But the government must acknowledge *now* that offsets are a serious problem and immediately begin to develop an effective framework for resolving it.

The United States requires a solid policy on offsets, not least because "every other serious aerospace nation has a coordinating body charged with nurturing and advancing domestic aerospace manufacturing, technology acquisition, and, of course, employment. The United States should do no less."[61]

How should we begin to formulate this policy? To begin with, the U.S. government must acknowledge the serious effects that offsets are having and will have on the aerospace industry. It must recognize the current utilization of offsets must be better understood.

A Commission

In order to gather information on offsets, a formal commission should be established by the President. The idea for such an entity is not new.[62] Several reputable studies have recommended it. Such an entity would bring together representatives from industry, labor, government, and academia to facilitate the gathering of information and to engage in meaningful dialogue over what can be done to establish a real policy on offsets—a policy which would promote the U.S. aerospace industry and its workers.

The commission would recommend policy and coordinate activities through efforts that would include a review of:

---

[60]The National Export Strategy Trade Promotion Coordinating Committee 5th Annual Report to the United States Congress, October 1997, p. 63.

[61]Barber and Scott, *Jobs on the Wing*, p. 78.

[62]See, e.g. *High Stakes Aviation, U.S. - Japan Technology Linkages in Transport Aircraft*, National Academy Press, Washington, D.C., 1994, p. 94; Barber and Scott, *Jobs on the Wing*, p. 3.

— transfer of jobs,
— transfer of technology,
— research and development,
— export sales and financing,
— review of license production and co-production agreements,
— subcontractor production,
— counter trade, and
— foreign investment.

The commission would also advise the Administration in negotiating relevant agreements and understandings with our trading partners. A priority should be given to negotiating restrictions on debilitating offsets that lead to arrangements that hurt U.S. aerospace workers. The commission could also develop a program to train and re-employ displaced aerospace workers. This program should allow workers who lose their jobs because of offsets to receive retraining and be eligible for trade adjustment assistance.

The commission could also develop concrete methods for facilitating the collection of data on offsets from both the public and private sectors. One approach is to require that any contractor directly or indirectly receiving federal monies identify and report specific information regarding offsets.

Importantly, this information should also be accessible to the public. The public has the right to know how its money is being spent. It has the right to know if its money is going to retain and create good jobs at home as or is being used to subsidize the creation of jobs in other countries. The public should also know if technology that was developed by their tax dollars is being transferred abroad.

CONCLUSION

Offsets create serious questions for policy makers concerned with the public interest. As discussed in this paper, offsets will have a growing negative impact on the lives of working Americans, particularly those whose livelihoods depend on the maintenance and expansion of the U.S. aerospace and related industries. Offsets can also have a very serious impact on the national security as valuable technology finds its way into other nations' defense-related activities. Furthermore, offsets have pit one group of private corporate interests against another as prime contractors sacrifice their relationships with sub-tier producers to satisfy offset arrangements.

Responsibility for creating a framework for resolving the issues that offsets raise lies with the U.S. government. Among other things, only the U.S. government has the resources and the authority for determining how we should proceed. Unfortunately, the U.S. government has yet to make any serious efforts in setting offset policy. While our government continues "dabbling" with the issue—meeting with interested parties on an ad hoc basis and encouraging "further discussion" on this topic—other nations have well-established policies on offsets and are moving rapidly toward utilizing their policies for their own benefits.

We can no longer idly sit by and relegate to U.S. private interests the sole responsibility of negotiating with other nations' governments over offset issues. To do so would be to abandon the role that government must play in protecting the public interest.

As the stakes get bigger and the pieces to the offset puzzle become more difficult to identify, we, as a nation, can no longer sit back and let other countries and the hundreds of private parties that are involved in the offset game set our course. It is time for the U.S. government to take a strong leadership role in developing our long over-due policy on offsets.

IV

ANNEXES

Defense Industry Offset Association Position on Offset Issues

Gordon Healey
Defense Industry Offset Association

BACKGROUND

The Defense Industry Offset Association (DIOA) is an organization of offset professionals from across the U.S. defense and aerospace industry. Its 65 member companies represent virtually 100 percent of the defense/aerospace prime contractors in the United States. Membership requirements specify that a member company be a U.S. corporation engaged in the defense business and that it have undertaken at least one offset obligation as a prime contractor to a foreign customer.

The purposes of the DIOA are to:

- educate its members about the practice of offsets and related business functions such as countertrade, joint venture formation, international finance, and transactional analysis;
- provide a forum and a means for its members to network with one another; and
- address business and policy issues affecting the practice of international offsets.

It is under this third objective that the DIOA sets forth the following position statement addressing issues raised about offsets.[1]

[1]The positions expressed below are those of the Defense Industry Offset Association (DIOA) as a professional organization and do not necessarily reflect the individual opinions of each of its constituent members. Gordon Healey, of Bell Helicopter, is serving as president of DIOA.

ISSUE: VOLUNTARY VERSUS MANDATORY OFFSETS

Industry Experience

Offsets are neither optional nor voluntary on the part of the U.S. defense/aerospace industry. They are *mandated*, either by the laws, regulations, or expectations of our customer countries; or they are *dictated* by competitive forces at work in the international marketplace. International sales are won on the basis of four fundamental criteria:

- price (low price or best value for the dollar),
- technical performance (how well does the product or system work?),
- offsets (industrial benefits provided to the customer country), and
- politics (regional favoritism, in-country political forces, buy-local preferences, etc.)

In sales where the competitors' prices and technical performance specifications are too close to call and where politics cannot be affected, offsets are often not only a requirement, but are actually the key to winning the sale. No U.S. aerospace company would voluntarily offer offsets to a customer. Offsets are expensive, difficult to manage, risky, time-consuming, politically unpalatable at home, and, in general, a nuisance.

The fact is, except in some unusual cases, U.S. aerospace companies cannot win international sales without offsets. To refuse the customer's offset requirement or expectation of offsets is to walk away from the sale and turn it over to the competitor.

ISSUE: THE IMPACT OF OFFSETS ON AEROSPACE EMPLOYMENT

Industry Experience

It will be rightly observed that in the course of fulfilling an offset obligation, a U.S. aerospace company may subcontract the manufacture of some of its components to foreign suppliers. It may also be asserted that this subcontract work typically has been or could be done by domestic workers or U.S. suppliers. Therefore—and this is a dangerous conclusion—the aerospace company is exporting U.S. jobs.

In fact, the DIOA finds that offsets are a significant factor *contributing* to aerospace employment in America. Not that the offset function itself employs so many people, but that offsets, being a key element to international sales, help ensure ongoing jobs to the thousands of U.S. aerospace workers who build the aircraft, vehicles, weapons, and systems that are sold to international customers.

The Aerospace Industries Association estimates that current total defense/aerospace employment in the United States is about 850,000 (not counting the U.S. military forces and civil servants). The companies that employ these 850,000 workers had combined sales in 1997 of $130 billion. Of that, about $50 billion was defense sales, $9.4 billion of which was exported to international customers. Using these figures, the proportions for work-force sizes can be easily calculated: Of the total 1997 work force, approximately 325,000 were in the defense sector, and 60,000 of these were employed in producing defense products for export. These workers rely on the success of international marketing and offset efforts for their jobs. If the data were available for *commercial* aerospace sales and offsets, it could be demonstrated that the figure of 60,000 jobs created or sustained as a result, in part, of offsets is actually much higher. In this paper, however, we focus only on the defense sector.

The fundamental flaw in the logic used by those who voice concern about offsets is this: The very jobs alleged to be exported by offsets are actually created and/or preserved by this practice. The tens of thousands of aerospace workers and suppliers who support the assembly lines for production of new F-15, F-16, and C-130 aircraft and the Apache, Cobra, and Blackhawk helicopters are now almost totally dependent on international sales for their livelihood. The international sales are won, in part, through effective offset commitments.

Are jobs exported in offset programs? Yes, a few, although "re-exported" might be a better term. Some of the work the aerospace prime contractors have traditionally done in-house is now placed in the manufacturing facilities of customer countries. Some of the subcontracts that have been traditionally bid solely to U.S. suppliers are now also bid to and won by suppliers in the customer countries. According to the U.S. Department of Commerce, to whom the aerospace companies annually report their offset activity, this amounts to about $3 billion worth of *offset credits.*

Understanding Offset Credits

It is important to understand that offset credits do not translate directly to dollars and hence to jobs. Due to mechanisms such as incentive multipliers, third-party joint venture formations, and offset projects that do not involve U.S.-based work, aerospace companies are often able to satisfy significant portions of their offset obligations without large expenditures and without impacting their labor base and supplier pool.

So what does $3 billion in offset credits mean in terms of actual defense/aerospace jobs? First, about half of the $3 billion was done as "indirect offsets," meaning mostly nondefense offset projects. Second, although no industry-wide research has been done on the effects of offset crediting mechanisms such as multipliers, one can safely suggest that at least half of the $1.5 billion in "direct" defense-related offset credits had little or nothing to do with actual jobs and subcontract work. Thus, of $3 billion in offset credits, only an estimated $500–$750 million would have been subcontract work. Returning to the aggregate figures, if the U.S. aerospace industry had $130 billion in sales and employed 850,000 workers, then how many workers would be associated with $500–$750 million in actual offset defense subcontracting? About 4,000.

In other words, for the 60,000 U.S. defense/aerospace workers whose jobs are sustained by international military sales, about 4,000 jobs are provided to our international customers under offset programs. It is not a bad exchange, particularly when one considers that if the U.S. defense companies had been prevented from engaging in offsets, they would have lost the sales which gave rise to the 60,000 jobs.

David Mowery in *Offsets in Commercial and Military Aerospace* characterizes the impact of offsets on aerospace employment as "minuscule." In the face of analyses such as *Jobs on the Wing*, Mowery states, "Overall, it is difficult to make a credible case that offsets in both military and commercial aerospace account for any but a small fraction of the sharp declines in aerospace employment since the 1980's." The DIOA agrees.

ISSUE: IMPACT OF OFFSETS ON THE SUBCONTRACTOR COMMUNITY

Industry Experience

Although the subcontractors sometimes feel like pawns in the offset process, the fact is that they are the beneficiaries of the prime contractors' international marketing successes. In the aggregate, 40 percent of the work that comes to a subcontractor is for export and probably has an offset obligation attached to it. The subcontractor may not be aware, for example, that his order for 50 shipsets of machined brackets for a pilot's seat will be delivered in the end to Italy and Taiwan, and that his prime contractor had to undertake offset obligations to win the contract.

Most second- and third-tier suppliers are never asked to participate in offset programs in the customer countries and thus are unwitting beneficiaries of the efforts of the offset managers to find other ways to offset that particular piece of the aircraft's value. From time to time it is necessary to give a purchase order to a firm in the customer country for a component that has been historically manufactured by a U.S. supplier. Such orders are placed, first, on the basis of competition, and second as "split orders" wherever possible to lessen the impact to the U.S. supplier and to retain the capability to manufacture the part in the United States. This is particularly true of critical, complex, or long-lead parts.

The DIOA notes that the U.S. Department of Commerce's annual report on *Offsets in Defense Trade* includes a section of anecdotes from suppliers who feel they have been hurt by offset programs from the primes. It should be pointed out that these anecdotal experiences are neither sustained nor verified by the 100+ preceding pages of empirical data collected by the Bureau for Export Administration. Furthermore, there has been no effort to determine the amount of work these suppliers received as a result of present or prior international sales by the prime contractors. There was, of course, no acknowledgment of benefit from the supplier when he received purchase orders in connection with such business, even though it was the prime contractor's offset program that had helped win the sales.

Addressing the Impact on Sub-Tier Suppliers

The DIOA acknowledges that the second- and third-tier suppliers are sometimes affected by offsets. Most of the time they are the beneficiaries of international sales. Occasionally, they lose work to foreign suppliers. The following suggestions are offered with respect to the sub-tier supplier situation:

- The Department of Commerce should conduct a more rigorous study of the effects of offsets on the suppliers, including the amount of work done by suppliers for eventual export.
- Suppliers who are losing contracts to foreign competitors should study their operations and take steps to become more productive and more competitive.
- For aerospace producers at all levels who have employees displaced by the internationalization of aerospace work, government programs to provide retraining and job placement should be improved and reemphasized.

ISSUE: LESSENING THE IMPACT OF OFFSETS

Industry Experience

Offset managers and aerospace contractors are not insensitive to the concerns of their work force and supplier base. On the contrary, there is universal appreciation for the obvious: Without its employees and its suppliers, a production company is dead.

The DIOA believes it is important for those affected by offset programs to understand the efforts an offset manager goes through to lessen the impact of offsets on the company's workers and supplier base.

Offset Contracts

In an offset agreement with a customer, the offset manager will negotiate for contract provisions that

- promise the lowest possible offset amount,
- offer the longest possible period of performance,

- provide for the widest possible spectrum of allowable projects,
- establish the most favorable crediting mechanisms for offset projects (including multipliers and lump sums for special projects), and
- authorize the longest possible list of eligible parties to participate in the offset effort.

Multipliers

The government offset administrators in many customer countries (Korea, Turkey, Australia, Singapore, Taiwan, Israel, Greece, Finland, and others) allow multipliers for offset projects meeting certain criteria. For example, an offset project directed to a key company in an economically depressed region of the country may receive $10 of offset credit for every $1 of project value. The U.S. aerospace company will focus on these highly incentivized project areas and, as a result, may reduce the offset effort to as little as one-tenth of the original obligation.

Technical Assistance and Training

Many international customers of U.S. aerospace products lack the technical know-how to operate and maintain the equipment they are buying. For offset projects, the manufacturers will often send technical specialists to the customers' facilities to perform training and assist the customers' technicians and operators in gaining much-needed expertise. The offset credits awarded for such projects usually are many times the expense incurred by the U.S. company.

Financing

Many customers of U.S. aerospace companies are the developing countries of the world whose defense budgets are small and who have limited experience in financing the acquisition of major aerospace or defense systems. The U.S. contractors will often provide assistance to their customers in approaching major Western investment banks or in structuring highly specialized financing schemes. The customer countries will award sizable amounts of offset credits for such assistance.

Market Development

Large aerospace companies frequently have a network of business contacts throughout the world that can be of assistance to firms in the customer countries for finding new export markets for their products. Millions of dollars of offset credits can be earned as such exports begin to flow from the customer countries.

Investments and Joint Ventures

This is a wonderful example of a win-win situation in the offset business. The U.S. aerospace company will find a U.S. firm who has a particular technology of interest to a firm in the customer country. The offset manager will then act as a marriage broker between the two firms, assisting in the formation of the joint venture, overseeing and sometimes participating in the investment, ensuring that technologies are properly licensed, and then collecting offset credits for the business generated by the joint venture. Such a business not only benefits the customer country but also opens up a new, international branch of the U.S. partner.

Non-U.S. Transactions

Many offset transactions involve the transfer of work from one international supplier to another. For example, an airplane manufacturer has landing gear components built in Israel for an offset obligation there. When the offset program is finished in Israel, the landing gear work is moved to the United Kingdom to be applied against a new offset commitment there. Such work transfers have no impact at all on U.S. jobs. Similarly, some countertrade transactions will involve the shipment of commercial goods between two non-U.S. countries, thus not affecting U.S. jobs nor creating increased competition at home.

Through these and other such creative methods, offset managers are able to greatly lessen the employment impact of offset obligations while still providing projects of significant value to the customer countries. Do offset administrators in the customer countries feel cheated when they see that they are only getting, for example, an aggregate of 4,000 jobs worth of aerospace work for their billions of dollars of expenditures on U.S. equipment? Not at all. It is they who provide the incentives and multipliers, and it is they who approve the offset projects presented to them by the U.S. offset managers. Although the system is a nuisance, it can work to the benefit of both parties.

ISSUE: TECHNOLOGY TRANSFER IN OFFSET PROGRAMS

Industry Experience

Technology transfer is, indeed, frequently an element of offset programs. It is highly desired by the customer countries and is almost always encouraged by large multipliers and incentives in terms of offset credits awarded. In this way, technology transfer projects tend to lessen the jobs impact of offset programs by substituting technology transferred for actual work transferred and provide a low-cost, mutually satisfactory means to fulfill an offset obligation.

The key issues in technology transfer projects are:

- Does the technology sent offshore in any way compromise U.S. security interests?
- Is the provider of the technology creating a foreign competitor?

U.S. Security

With respect to the impact on U.S. security, the DIOA points out that the U.S. government operates a sophisticated export controls mechanism through which all requests to export technology must pass. Data export licenses and technical assistance agreements are issued by the U.S. government prior to the delivery of any aerospace technology into the hands of a foreign entity. The DIOA is persuaded that this mechanism is working well and that those who have concerns with the sensitivity of data exported under offset programs (or any international joint venture) should address their concerns to the U.S. State Department.

Creating Competitors

Concerning the creation of foreign competitors through technology transfers, views are more subjective. One must ask, however, if an aerospace company gives away a core technology, in the interest of fulfilling an offset program, does it thus create a competitor for itself? We believe that companies are not that short-sighted. And if the company is so short-sighted, it deserves to be put out of business by the foreign firms it helps establish. Moreover, it should be noted that, even with technology transfers, the cost to enter and remain in today's aerospace industry is prohibitive. The trend for aerospace companies has been exit, not entry.

Kenneth Flamm, in *The Policy Context for Military Aerospace Offsets*, addresses at some length a concern that the waiving of research and development (R&D) recoupment costs in technology transfers is, in effect, selling the U.S. taxpayer short. The DIOA would argue that the U.S. taxpayers' primary payoffs for their R&D expenditures come as the aircraft or defense systems so developed are delivered to the U.S. forces. Any R&D recoupments paid by foreign concerns are simply additional revenue for the U.S. government. Where such recoupments are waived in the interest of providing an incentive for the foreign government to buy American, the U.S. taxpayer still benefits because of the economies of scale that arise from the increased quantities sold offshore. The Department of Defense's unit price for defense systems goes down because costs are spread over a greater number of units.

DIOA RECOMMENDATIONS

International offsets are anything but a perfect business practice, and as such can be improved. In the best of worlds, they can even be done away with. But we do not foresee "some grand agreement" as a near-term solution to offsets or to any aspect of the practice. Solutions will evolve with the changing nature of international business. The DIOA offers the following recommendations to guide the process:

1. Take no unilateral action. Offsets are an established element of international trade. Our foreign trading partners will continue to demand them, and our international competitors will continue to offer them. Consequently, any unilateral action on the part of the United States to prevent its companies from engaging in offsets would be highly detrimental to the international business interests of the U.S. defense/aerospace industry.

2. Improve the data collection process. The DIOA agrees that regular reports to the U.S. government on offset performance can be helpful to any who want to understand the magnitude and impacts of this unusual business practice. We suggest, however, that improvements could be made in the data collection process and would offer our assistance to the U.S. Department of Commerce in modifying its survey questionnaires and report forms so as to gather more accurate, meaningful offset data.

3. Continue with multilateral discussions. The DIOA welcomes any *mutual* concessions that could be made between the United States and its trading partners that would limit the practice of offsets without imperiling the international marketing process.

4. Maintain dialogue. As long as the various parties who are interested in offsets—the prime contractors, the subcontractors, the labor unions, the government, etc.—are talking to each other, there is hope for progress toward a consensus. And if, in the unhappy event that a consensus is not within reach, there is still value in understanding one another's positions. The DIOA will continue to support efforts to maintain a dialog on offset issues.

5. Keep offsets in perspective. In terms of the aggregate U.S. aerospace business and the national trade balance, the impact of offsets is small. True, offsets are sometimes highly influential in securing international sales, but the ebb and flow of aerospace jobs and technologies due to specific offset transactions are dwarfed by other factors such as defense downsizing and industry restructuring. The U.S. government has wisely adopted a "hands off" national offset policy, and unless or until significant harm—or even the risk of significant harm—can be demonstrated, we will do well to leave this powerful marketing tool in the hands of the defense/aerospace firms where it can be used to secure jobs for their employees and suppliers.

Participants

Beth Almeida
University of Massachusetts, Amherst

Stephen Austin
Office of the Deputy Under Secretary
U.S. Department of Defense

William Balhaus
Office of Science and Engineering (Ret.)
Lockheed Martin Corporation

Randy Barber
Center for Economic Organizing

Sally Bath
Office of Aerospace
U.S. Department of Commerce

Steve Beckman
United Auto Workers

Lawrence Bertino
Boeing Space Systems

Brad Botwin
Strategic Industries & Economic Security
U.S. Department of Commerce

Kirk Bozdogan
Lean Aerospace Initiative
Massachusetts Institute of Technology

Don Brown
Boeing

H. Lee Buchanan
DARPA

Steven C. Clemons
Economic Strategy Institute

Wesley Cohen
Massachusetts Institute of Technology

E. William Colglazier
Executive Officer
National Research Council

Bill Cosper
General Dynamics

Mark Crawford
New Technology Week

R. J. Donovan
Trade Promotion Coordinating Committee
U.S. Department of Commerce

Donald Eiss
Office of the United States Trade Representative

James A. Falco
Massachusetts Institute of Technology

Kenneth Flamm
The Brookings Institution

Tom Flowers
Offset and Business Management
Northrop Grumman International, Inc.

T. J. Glauthier
Office of Management and Budget

Gordon Healey
International Offsets and Counter Trade
Bell Helicopter Textron, Inc.

Owen Herrnstadt
Intl. Assoc. of Machinists and Aerospace Workers

Page Hoeper
International and Commercial Programs
U.S. Department of Defense

Art Ismay
Rockwell International Trading Company

Kenneth Jarboe
Jarboe and Associates

Jeri Jensen-Moran
Trade Promotion Coordinating Committee
U.S. Department of Commerce

Joel Johnson
Aerospace Industries Association

Albert Kelley
Department of Aeronautics and Astronautics
Massachusetts Institute of Technology

Lena Lawrence
STEP Board
National Research Council

Thea Lee
Public Policy Department
AFL-CIO

Greg Martin
Corporation Industrial Participation
Lockheed Martin Corporation

Paul McNeill
International Commercial Programs
Boeing Corporation

Stephen A. Merrill
STEP Board
National Research Council

Martin A. Meth
Industrial Capabilities and Assessments
U.S. Department of Defense

David C. Mowery
Haas School of Business
University of California, Berkeley

Deborah Nightingale
Lean Aerospace Initiatiave
Massachusetts Institute of Technology

John Oldfield
STEP Board
National Research Council

Bartley P. Osborne, Jr.

Frank Parker
International Marketing
ITT Defense & Electronics

Gary Powell
Industrial Capabilities and Assessments
U.S. Department of Defense

William Reinsch
Bureau of Export Administration
U.S. Department of Commerce

Richard Ridge
General Electric Engines
General Electric

Dorothy Robyn
National Economic Council
The White House

Howard Rosen
Joint Economic Committee
U.S. Congress

John Sandford
Rolls Royce, N.A.

Robert E. Scott
The Economic Policy Institute

Barbara Shailor
International Affairs Department
AFL-CIO

John A. Shaw
Cambridge Consulting Group

Steve Sleigh
Strategic Resources
International Association of Machinists

George Souteropoulos
General Electric

Captain Carlos Souto
Embassy of Portugal

John Tucker
Strategic Industries & Economic Security
U.S. Department of Commerce

Myron Uman
Acting Director, Policy Division
National Research Council

Al Volkman
International and Commercial Programs
U.S. Department of Defense

Todd Watkins
College of Business and Economics
Lehigh University

David Welch
Industrial Programs
Lockheed Martin Tactical Aircraft Systems

Kimberly Wells
Office of Aerospace
International Trade Administration

Charles W. Wessner
STEP Board
National Research Council

Marvin Winkelmann
International and Commercial Programs
U.S. Department of Defense

Alan Wm. Wolff
STEP Board and
Dewey Ballantine

Joel Yudken
AFL-CIO

Selected Bibliography

Barber, Randy and Robert E. Scott, *Jobs on the Wing: Trading Away the Future of the U.S. Aerospace Industry.* Economic Policy Institute, Washington, D.C., 1995

Bracken, Paul, *Non-Standard Models of the Diffusion of Military Technologies: An Alternative View.* Technical report prepared for the Director of the Office of the Secretary of Defense, Net Assessment. Joint Management Services, Dunwoody, Georgia, February 1997

Cole, Jeff, "Report Assails Defense-Sector 'Offset' Deals," *Wall Street Journal,* June 22, 1994

Defense Science Board Task Force on *Vertical Integration and Supplier Decisions.* Office of the Secretary of Defense, Department of Defense, Washington, D.C., 1997

Department of Commerce, *Offsets in Defense Trade: A Study Conducted Under Section 309 of the Defense Production Act of 1950, As Amended.* Bureau of Export Administration, Washington, D.C., August, 1997.

Department of Commerce, *Offsets in Defense Trade: A Study Conducted Under Section 309 of the Defense Production Act of 1950, As Amended.* Bureau of Export Administration, Washington, D.C., May 1996.

Eisenhour, John H., *Offset: The Political Dimension of Countertrade,* unpublished manuscript, 1994

European Commission. 1994 and 1997. *The European Aerospace Industry: Trading Position and Figures.* Brussels, Belgium: European Commission, Directorate-General III.

"Federal Funding of Research Development in Transportation: The Case of Aviation," Presented at the Symposium on the Effects of Federal R&D, National Academy of Sciences, Nov. 19-20, 1985. *The Effects of Federal Research Funding.* National Academy Press, Washington, D.C., 1986

General Accounting Office, *Asian Aeronautics: Technology Acquisition Drives Industry Development,* NSIAD-94-140, Washington, D.C., May 1994

— *International Trade: Long-Term Viability of U.S.-European Aircraft Agreement Uncertain,* GGD-95-45, Washington, D.C., December 1994

— *Military Exports: Concerns Over Offsets Generated with U.S. Foreign Military Financing Program Funds,* NSIAD-94-127, Washington, D.C., June 1994

— *Military Exports: Offset Demands Continue to Grow.* GAO/NSIAD-96-65, Washington, D.C., April 1996

— *Export Controls: Sensitive Machine Tools Exports to China,* November, 1996.

— *Military Sales: Concerns Over Offsets Generated Using U.S. Foreign Military Financing Program Funds (Testimony),* T-NSIAD-94-215, Washington, D.C., June 1994

— *Military Offsets: Regulations Needed to Implement Prohibition on Incentive Payments*, GAO/NSIAD-97-189, Washington, D.C. August, 1997.

Greider, William, *One World, Ready or Not: The Manic Logic of Global Capitalism.* Simon & Schuster, New York, N.Y., 1997

Johnson, Joel L., "The United States: Partnership with Europe," in Ethan B. Kapstein, ed., *Global Arms Production: Policy Dilemmas for the 1990s.* University Press of America, Lanham, Maryland, 1992

Louscher, David J., *The Competitive Environment for Future International Defense Systems Sales, Vol. I.* Foresight International, Bath, Ohio, April 1997

Louscher, David J. and Michael D. Salomone, *Marketing Security Assistance: New Perspectives on Arms Sales.* Lexington Books, Lexington, Mass., 1987

Lumpe, Lora, "Sweet Deals, Stolen Jobs." *The Bulletin of the Atomic Scientists,* September/ October 1994, pp. 30-35

Manufacturers Alliance, *Offsets in Foreign Sales of Defense and Nondefense Equipment: A Manufacturers Alliance Review.* Arlington, VA, February 1997

Martin, Stephen, *The Economics of Offsets: Defense Procurement and Countertrade.* Harwood Academic Publishers, The Netherlands, 1996

Maskus, Keith E. and Denise R. Eby, "Developing New Rules and Disciplines on Trade-Related Investment Measures," in Robert M. Stern, ed., *The Multilateral Trading System: Analysis and Options for Change.* University of Michigan Press, Ann Arbor, 1993

Marvel, K. Barry, "International Offsets: An International Development Tool," Contract Management Vol. 4, October, 1995.

Mastel, Greg, *The Rise of the Chinese Economy: The Middle Kingdom Emerges.* M.E. Sharpe, Armonk, N.Y., 1997

Mintz, John, "GAO Report Critical of Military 'Offsets,'" *The Washington Post* June 22, 1994

Mowery, David C., *Alliance Politics and Economics: Multinational Joint Ventures in Commercial Aircraft.* Ballinger, Cambridge, Mass., 1987

Mowery, David C., ed., *International Collaborative Ventures in U.S. Manufacturing.* Ballinger, Cambridge, Mass., 1988

Mowery, David C. and Nathan Rosenburg, "Government Policy and Innovation in the Commercial Aircraft Industry, 1925-75." In R.R. Nelson, ed., *Government and Technical Change: A Cross-Industry Analysis.* Pergamon, New York, N.Y., 1982

— "Commercial Aircraft: Cooperation Between the U.S. and Japan." *California Management Review,* 1985, pp. 70-92

National Export Strategy Trade Promotion Coordinating Committee 5th Annual Report to Congress, October, 1997.

National Research Council, *Conflict and Cooperation in National Competition for High-Technology Industry.* National Academy Press, Washington, D.C., 1996

— *High-Stakes Aviation: U.S.-Japan Technology Linkages in Transport Aircraft.* National Academy Press, Washington, D.C., 1994

Office of Management and Budget, *Impact of Offsets in Defense-Related Exports: A Summary of the First Three Annual Reports.* U.S. Executive Office of the President, Washington, D.C., December 1987

— *Offsets in Military Exports.* U.S. Executive Office of the President, Washington, D.C., July 16, 1990

— *Second Annual Report on the Impact of Offsets in Defense-Related Exports.* U.S. Executive Office of the President, Washington, D.C., December 1986

Office of Management and Budget. 1990. *Study of Military Offsets.* Washington, D.C.: U.S. Office of Management and Budget.

Office of Technology Assessment, *Global Arms Trade: Commerce in Advanced Military Technology and Weapons.* Congress of the United States, Washington, D.C., 1991

"Partnership: The Price of Entry." *Aerospace America,* June 1996, pp. 29-32

Samuels, Richard, *Rich Nation, Strong Army: National Security and the Technological Transformation of Japan.* Cornell University Press, Ithaca, N.Y., 1994.

Scott, Robert E. and Randy Barber, "Aircraft: Offsets Eroding U.S. Technological Lead," *New Technology Week,* October 15, 1996, pp. 5-6.

Sperling, James, David Louscher, and Michael Salomone, "A Reconceptualization of the Arms Transfer Problem." *Defense Analysis,* vol. 11, No. 3, 1995, pp. 293-311

"Tools of the Trade: Offsets, Outsourcing, and Joint Ventures." *Aerospace America,* September 1996, pp. 10-12

Trade Promotion Coordinating Committee, *National Export Strategy: Toward the Next American Century: A U.S. Strategic Response to Foreign Competitive Practices.* U.S. Government Printing Office, Washington, D.C., October 1996

Tyson, Laura, *Who's Bashing Whom: Trade Conflict in High-Technology Industries.* Institute for International Economics, Washington, D.C., 1992

Udis, Bernard and Keith E. Maskus, "Offsets as Industrial Policy: Lessons from Aerospace," *Defense Economics,* vol. 2, 1991, pp. 151-164

U.S. Congress, House of Representatives, Committee on Government Reform and Oversight, "Foreign Offset Demands in Defense and Civil Aerospace Transactions," October 23, 1998, Washington, D.C.

Wessner, Charles W. and Alan Wm. Wolff, eds. *Policy Issues in Aerospace Offsets: Report of a Workshop.* National Academy Press, Washington, D.C. 1997.

Womack, J.P. and D.T. Jones, "From lean production to lean enterprise." *Harvard Business Review.* March-April, 1994.